Multiple Action-Based Design Approaches to Antibacterials

John Bremner

Multiple Action-Based Design Approaches to Antibacterials

Springer

John Bremner
School of Chemistry and Molecular Bioscience
University of Wollongong
Wollongong, NSW, Australia

Illawarra Health and Medical Research Institute
Wollongong, NSW, Australia

ISBN 978-981-16-0998-5 ISBN 978-981-16-0999-2 (eBook)
https://doi.org/10.1007/978-981-16-0999-2

This Springer imprint is published by the registered company Springer Nature Singapore Pte Ltd.
The registered company address is: 152 Beach Road, #21-01/04 Gateway East, Singapore 189721, Singapore

'Ancora imparo' I am still learning.

—Michelangelo

Preface

The main rationale for this book was to bring together the intentional design and development aspects of small molecules with actual or potential synergistic multiple modes of action in the context of antibacterial results. With the ever increasing health threat from multi-drug-resistant human pathogenic bacteria, this is a pressing issue for researchers. The stress in this book is on design principles and ideas rather than a comprehensive review compilation. There is some coverage of dual activity approaches in the book, but this area has been well reviewed previously so dual activity design ideas are used mainly as a springboard for higher-order design guidelines. Thus, the book highlights known and possible ways to achieve triple or higher activities, including activities which are directly antibacterial or which indirectly assist such activity, and focussing on medicinal chemistry aspects. The book covers underlying design aspects for combinations of drugs, single molecule hybrids with actual or potential multiple actions, prodrugs which could provide access in situ to multiple interactions, and future design possibilities based mainly on new activity pathways. Interdisciplinary aspects are also covered in sufficient depth to adequately inform the medicinal chemistry design.

Writing a book is like climbing a mountain. Sometimes, though, even the base camp seems beyond reach, but hopefully this book will help to suitably position readers at this camp to explore the many different possible routes to the summit.

The photographs at the start of Chaps. 2 and 4, and the images in Figs. 1.2 and 1.3 were purchased from Shutterstock, and I thank Dr. David Rhodes for the photograph heading Chap. 1 and Mrs. Susan Bremner for those at the beginning of Chaps. 3 and 5. I also thank the editor of the Srinakharinwirot University Science Journal for permission to use a Scheme and Figures from Bremner 2017 (SWU Sc J 33(1):1–19) as Scheme 4.1 and Figs. 3.5, 3.12, 3.19, 3.22, 3.28–3.30a, b and 3.32–3.33 in this book.

I would like to gratefully acknowledge students and fellow staff at the University of Wollongong for their help through discussions and collaborative research over many years which has been such an inspiration and encouragement. I would also like to sincerely acknowledge my past teachers and mentors, and students and colleagues at the University of Tasmania and many other institutions both in Australia and abroad, who have contributed so much. Invaluable assistance from Associate

Professor Siritron Samosorn, Srinakharinwirot University, Bangkok, Thailand, with the drawing of structures and referencing is also very gratefully acknowledged. The author would also like to thank the School of Chemistry and Molecular Bioscience, the former Centre for Medical and Molecular Bioscience, Molecular Horizons, the Illawarra Health and Medical Research Institute, and the University of Wollongong for on-going support and assistance, and staff at Springer for their encouragement and help. Finally, my heartfelt thanks to my wife, Susan, without whom this book would not have been possible.

Wollongong, Australia
January 2021

John Bremner

Contents

About the Author

John Bremner was born in Perth, Western Australia, and after completion of his undergraduate studies at the University of Western Australia, he undertook a Ph.D. at the Australian National University. After a further year at Harvard University as a Research Fellow, he returned to Australia to take up a Lectureship in Chemistry at the University of Tasmania. Later, on sabbatical leave, he completed a Diploma in Chemical Pharmacology at the University of Edinburgh. In 1991, he was appointed as Professor of Organic Chemistry at the University of Wollongong in NSW, Australia. He retired in 2007 and is now an Emeritus Professor of the University of Wollongong. He is a Fellow of the Royal Australian Chemical Institute and received a Distinguished Fellow Award of the RACI in 2011. In 2001, he received the Adrien Albert Award in medicinal chemistry. He is also a Fellow of the Royal Society of Chemistry.

Chapter 1
Antibacterials

> *"What you see is that the most outstanding feature of life's history is a constant domination by bacteria."*
> —Stephen Jay Gould, American evolutionary biologist, palaeontologist, and science historian; 1941–2002.

Abstract This chapter sets the basic perspectives for the other chapters in the book. In this context, aspects discussed include the ever-rising global health problem from drug resistant pathogenic bacteria, a basic review of bacterial resistance mechanisms, and the need for innovative solutions to overcome resistance to antibacterials. General ways to achieve multiple-action effects through drug combinations, one or more being directly antibacterial, and through single molecule hybrids and prodrugs, are introduced. Relevant definitions are also outlined in Sect. 1.5 and the chapter concludes with an introduction to approaches towards achieving bacterial over host selectivity.

1.1 Bacteria

Bacteria, unicellular microorganisms which lack membrane bound organelles (prokaryotes), have been remarkably successful in evolutionary terms colonising most environments on earth. Generally it is thought they developed separately from the archaea, microorganisms which thrive in more extreme environments, and from

J. Bremner, *Multiple Action-Based Design Approaches to Antibacterials*,
https://doi.org/10.1007/978-981-16-0999-2_1

the eukaryotes although certain early evolutionary aspects of the derivation of the last two remain contentious (Watson 2019).

Traditionally, bacteria are categorised by how they react to staining with the Gram stain, Gentian Violet, followed by potassium triodide as a post-mordant. Hans Christian Gram was a Danish bacteriologist who invented the staining procedure. Gram-positive bacteria retain the dye after an ethanol wash, while bacteria designated as Gram-negative do not retain this purple dye but are identified by retention of a subsequently added dye, Safranine, with the bacteria being stained red (Stinson 1996). The differential dyeing characteristics reflect cell wall differences with Gram-positive bacteria having thick cross-linked polysaccharide cell walls which adsorb the gentian violet strongly. On the other hand, the Gram-negative bacteria have thin polysaccharide cell walls covered by a negatively-charged phospholipid layer for adsorption of the positively charged Safranine. Other bacterial groups are comprised of the non-stain reacting bacteria and the Mycobacteria.

Pathogenic and non-pathogenic bacteria can come from any of these different groups above, for example the pathogenic bacterium golden staph., *Staphylococcus aureus* (Gram-positive) and *Pseudomonas aeruginosa* (Gram-negative). The pathogenic *Mycobacterium tuberculosis*, the cause of tuberculosis, is neither Gram-positive nor Gram-negative due to a waxy coating on the cell surface (mainly mycolic acid) which obviates staining. *Mycobacterium tuberculosis* can be stained by an alternative staining procedure, however, and is classified as an acid-fast Gram-positive due the lack of an outer cell wall membrane.

Resistance by human pathogenic bacteria to previously effective antibacterial treatments poses an ever increasing threat to human health worldwide, a threat which could be exacerbated by global climate change with warming temperatures (Blair 2018; MacFadden et al. 2018). While a variety of antibacterials have been used over the years, the vigorous counter attack by bacteria continues as emphasised in some recent reviews (World Health Organization 2017a; The Review on Antimicrobial Resistance 2016). In early 2017, the World Health Organization published, for the first time, a ranked threat list of bacteria or bacterial families that posed the most pressing risk to human health and for which new antibacterials were needed (World Health Organization 2017b; and comments by Willyard 2017). Significantly the top three bacteria/families were in the Gram-negative category with the antibiotic type to which they are resistant noted in parentheses: *Acinetobacter baumanii* (carbapenem), *Pseudomonas aeruginosa* (carbapenem), and ESBL (extended spectrum-β-lactamase)-producing *Enterobacteriaceae* (carbapenem; 3rd generation cephalosporin). Watkins and Plattner (2015) have written a very good retrospective on antibacterials, including a summary of the carbapenems as well as covering the generations of cephalosporins. While not on the WHO ranked threat list, the universally distributed Gram-positive *Staphylococcus epidermidis*, an important cause of hospital-acquired infections, should now also be considered following the discovery that some strains of this pathogen revealed in Europe are resistant to all known antibiotics (Lee et al. 2018). They are thus classified as MDR (Multi Drug Resistant) strains and are hence of significant concern.

Tackling the antibacterial resistance problem presents key multi-disciplinary research challenges across the fundamental, preclinical and clinical sciences. The development of new antibacterials to treat Gram-negative pathogen-mediated disease is particularly critical, as there is only a relatively small number of drugs approved or in the pipeline which could potentially treat infections resulting from multidrug resistant Gram-negative ESKAPE bacterial pathogens. The ESKAPE pathogens include *Enterococcus faecium*, *Staphylococcus aureus*, *Klebsiella pneumoniae*, *Acinetobacter baumanii*, *Pseudomonas aeruginosa*, and *Enterobacter* spp., the last four of which are Gram-negative. This weakness in the armoury against human pathogenic Gram-negatives has been noted in the excellent clinical antibiotic pipeline update in the review by Butler et al. (2017) as well as in another review by Koulenti et al. (2019), focussing on new drugs and potential future ones active against multidrug-resistant Gram-negative bacteria (Koulenti et al. 2019). Further helpful detail is also provided in the article by Breijyeh et al. (2020) including approaches to try and resolve the issue with resistant Gram-negatives. The seriousness of this situation cannot be overemphasized especially as in some cases virtually no effective antibacterials are available. While the issue needs to be assessed at a global level, therapeutic efficacy differences also appear at country, community and individual levels, the last as illustrated by a tragic fatal case of infection with cPan-drug resistant *Klebsiella pneumoniae*: an isolate of this bacterium was resistant to 26 antibiotics (Chen et al. 2017).

Although neither Gram-negative nor Gram-positive, extremely drug resistant (XDR) *Mycobacterium tuberculosis* is also a great health concern, and specifics on antibacterial development possibilities for tuberculosis are covered in Mishra et al. (2017) including clinical manifestations of the disease, existing therapies and drugs in the pipeline, molecular targets for antimycobacterial agents, assay techniques, and alkaloids with activity against this pernicious mycobacterium. A further detailed coverage of all aspects of this major disease burden on global health, including recent drug developments and drug trials, is given in the annual WHO reports (World Health Organisation 2018, 2019a).

Antibacterial agents in the clinical development pipeline have been well analysed by a number of groups including by Theuretzbacher and colleagues (Theuretzbacher et al. 2019, 2020a), Butler and Paterson (2020), and the World Health Organisation (2017a, 2019b). A useful report by The Pew Charitable Trusts (2016) highlighted the pressing need to increase the number of new antibacterial drugs in the pipeline in a holistic overview of the scientific priorities involved and ways to try and meet them. These Trusts also provide regular updates in this area and as of December 2019 around 41 new antibiotics with the potential to treat serious infections caused by bacteria in global clinical development (The Pew Charitable Trusts 2020). Agents in the preclinical phase have also been analysed by Theuretzbacher et al. (2020b) and the World Health Organisation (2019c). While it is encouraging to have a number of compounds in the preclinical pipeline there is still an urgent need for more, particularly from different structural types and with new modes of action.

In view of the multi-dimensional nature of the bacterial disease challenge, various treatments have been, and are being, pursued. These include preventative hygiene measures and vaccines, as well as biological and chemical therapies based on viruses or small molecule antibacterials or antibiotics. Strictly speaking antibiotics are substances produced by microorganisms that compromise the survival of other microorganisms generally, while an antibacterial negatively impacts, either directly or indirectly, on the growth or survival of bacteria. Antibacterials can thus be either naturally derived or accessible through total laboratory synthesis, or semi-synthesis involving modification of a natural product precursor. The broader term 'antibacterial' is thus used for the most part in this book rather than the term 'antibiotic' although, arguably, the division between natural vs synthetic (or semi-synthetic) is a somewhat tenuous one. Humans are intrinsically part of nature not apart from it, and in that sense 'synthetic' is also 'natural' since the synthesis of new antibacterials in laboratories involves humans or human-made or designed automated processes.

Not consciously making the natural-synthetic division can help to inform antibacterial design in interesting new ways. For example, the recognition of the human microbiome itself as a producer of antimicrobial agents, mainly peptidic antimicrobials, is opening up new possibilities for design and synthesis and potentially in vivo manipulation of such agents. One wonders also about non-peptidic small molecules being produced as direct antibacterials or as a part of combinations which might include activity against resistance mechanisms within the human microbiome complex and/or in human tissues as part of a supplementary defense protocol against bacterial pathogens (Garcia-Gutierrez et al. 2019).

Streptomyces, a bacterial genus known for producing antibiotics, has been identified in the human gut microbiome (Seipke et al. 2012; Bolourian and Mojtahedi 2018) and in association with human lung tissue for example, where a particular *Streptomyces* sp. TR1341 was isolated and the secondary metabolites produced were implicated in a range of bioactivities including antibacterial activity against some Gram-positive and Gram-negative bacterial pathogens (Herbrík et al. 2020). Additionally genomes for cyanobacteria have also been detected in the human gut microbiota (Almeida et al. 2019), one such being from a non-photosynthetic cyanobacterial-related clade known as Melainabacteria. Photosynthetic cyanobacteria are known to produce a range of biologically active peptidic and non-peptidic metabolites including some with antibacterial activity (Dixit and Suseela 2013), for example the hapalindoles from two cultured cyanobacteria (Kim et al. 2012). The structure of hapalindole A is shown in Fig. 1.1a and the related indole-fused isonitrile compound ambiguine I isonitrile (Raveh and Carmeli 2007), from another species of cyanobacteria, in Fig. 1.1b. Both hapalindole A and ambiguine I isonitrile show good antibacterial activity and hapalindole A is very potent against *Mycobacterium tuberculosum*. Perhaps the non-photosynthetic relatives could also produce antibacterial metabolites and possibly with multi-targeting actions. By thinking outside the 'natural' or 'biological' square as such and by considering a more global natural-synthetic square, new design possibilities are likely to emerge, including advances from looking more at how best to work with nature and enhance it.

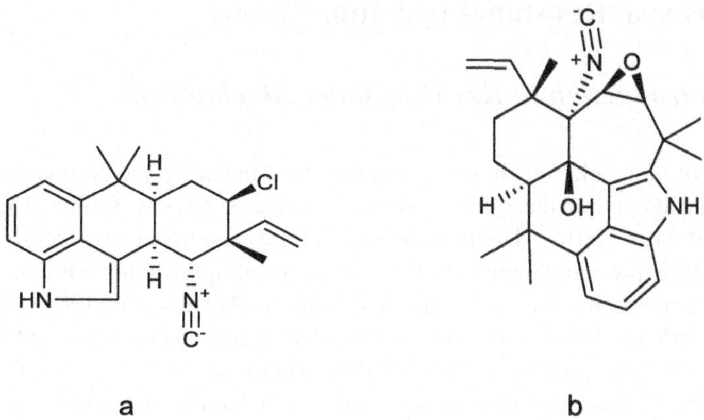

Fig. 1.1 Structure of hapalindole A (**a**) and ambiguine I isonitrile (**b**)

1.2 Antibacterial Action

Antibacterial action can be mediated by both direct and indirect means and result in bactericidal or bacteriostatic outcomes over various time courses. Diverse targets in bacterial cells for direct actions involve cell wall synthesis and maintenance, the biomembrane, protein synthesis, enzyme inhibition and interference with DNA synthesis, replication and function. Indirect actions can encompass inhibition of protein-based efflux pumps or other transporters, or suppression of external tagets like quorum sensing agents or virulence factors.

As well as these actions, 'ideal' antibacterials need to have a low or zero rate of bacterial resistance development, and be bacterially selective as much as possible in order to minimise toxic effects on host cells as well as being selectively toxic for the pathogenic bacteria versus good bacteria in the intestinal microbiome if orally administered. If applied topically, effects on the skin microbiome also need to be considered. No cross-resistance characteristics should be evident either, while favourable ADME (absorption, distribution, metabolism and excretion) properties, no teratogenic effects and no adverse drug-drug interactions with combination treatments are also important considerations. The characteristics of a so-called 'ideal antibacterial' have been summarised and discussed (Singh et al. 2017; Gajdács 2019) and these considerations are important in the design process for new multi-action antibacterials even if all the characteristics can only be partially achieved. The key point is that the design and development process is a necessarily complex one with an interacting matrix of factors to be taken into account. The discovery of new antibiotics or antibacterials has rightly been referred to (Lewis 2020) as a 'science' and one with many disciplinary inputs.

1.3 Bacterial Resistance to Antibacterials

1.3.1 Introduction to Key Resistance Mechanisms

A number of molecular mechanisms have evolved in bacteria to combat exposure to antibacterial agents and these are covered in detailed reviews by Blair et al. (2015) and Munita and Arias (2016). The mechanisms fall into a number of general categories including the expression of multiple types of efflux pumps to export antibacterials and reduce their intracellular concentrations to sub-lethal levels; alteration of biological target structures or increased expression of such targets; changes in metabolism and metabolic activity; sporulation; and biofilm formation.

Biologically, the types of resistance are characterised as intrinsic, acquired or adaptive. Intrinsic resistance is defined as the inherent ability of an individual species of bacterium, as a result of its structural or functional properties, to resist the action of an antibiotic or antibacterial agent. This may occur because of a lack of the target of the antibiotic or through barriers to permeation for example. In contrast, acquired resistance involves the bacterium picking up or acquiring resistance elements for counteracting particular antibiotics. These elements may include ways to minimize the intracellular concentration of the antibiotic, modification of the antibiotic target site, or chemical modification of the antibiotic to negate its activity (Blair et al. 2014).

Adaptive resistance is characterised by the temporary acquisition of resistance elements in response to external threats as could be the case on exposure to an antibiotic (Gorityala et al. 2016). In nature, antibiotics have a number of different roles and the way antibiotic resistance is developed in the natural environment is important when considering which mechanisms transfer to human pathogenic bacteria. Natural antibiotics or antibacterials from non-human sources are thought to have evolved to control competitive bacteria in the soil and aquatic environment. The human microbiome does, however, produce antimicrobial agents for bacterial competitors and these can display activity against some human pathogenic bacteria like *Clostridium difficile*, which is now named as *Clostridioides difficile* as noted by the Centers for Disease Control and Prevention in the USA (CDC 2019). Both taxonomic names for this species are used in this book reflecting the name used in the original sources. For a comprehensive multi-authored set of reviews in this general resistance area see Walsh (2015).

New agents which can block or attenuate antibiotic resistance would be of great interest in the fight against bacterial resistance. Recent work on slowing such antibiotic resistance by an anti-oxidant (Edaravone) in the presence of the antibacterial ciprofloxacin in pathogenic *Escherichia coli* is of considerable significance in this connection and points to the possible further development of small molecules to block the evolutionary development of resistance (anti-evolvability drugs) (Pribis et al. 2019).

1.3.2 *Resistance and Resilience*

It is also important here to differentiate between resistance and resilience since it needs to be considered in new antibacterial design parameters. Resistance can be defined as the lack of response by bacteria on treatment with antibacterial drugs. In contrast to resistance, resilience involves the bacterial community and reflects the capacity of the community or system to recover its original state after disruption. The term antibiotic resilience is now suggested when antibiotics cause the dislocation or disturbance, and the microbiological indicators that can be considered to measure this and assess the anticipated capacity of bacteria to recover from treatments with antibiotics have also been proposed (Carvalho et al. 2019).

1.3.3 *The Gram-Negative Challenge*

The development of new antibacterials to treat Gram-negative pathogen-induced disease is particularly critical as emphasised by Butler et al. (2017) in their review and also by Tacconelli et al. (2018). Such bacteria are inherently difficult to counter partly as a result of their multi-layered cell envelope (Fig. 1.2) which physically hinders antibacterial ingress. With Gram-negative bacteria there is an outer membrane decorated with externally projecting lipopolysaccharide (LPS) chains and containing proteins like the porins, a central peptidoglycan layer, and then the cytoplasmic

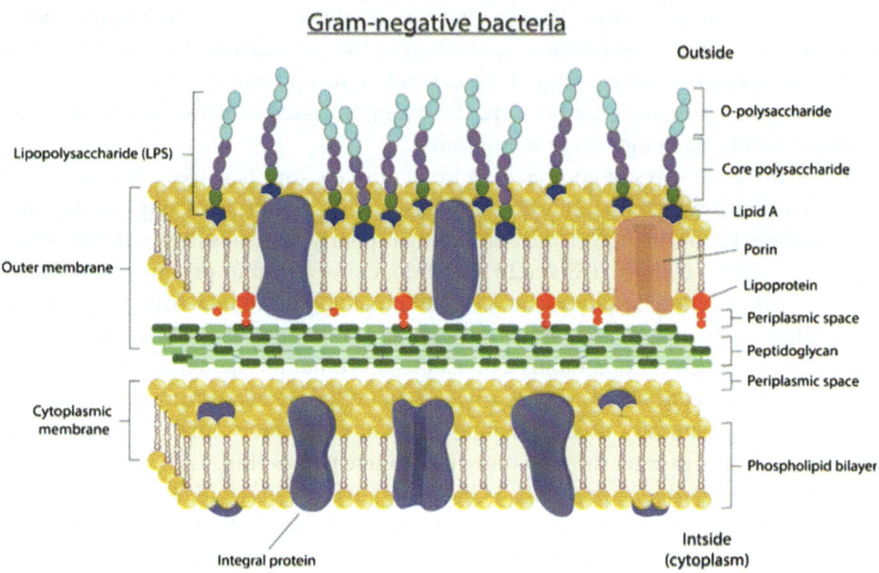

Fig. 1.2 Model of the cell wall structure in Gram-negative bacteria

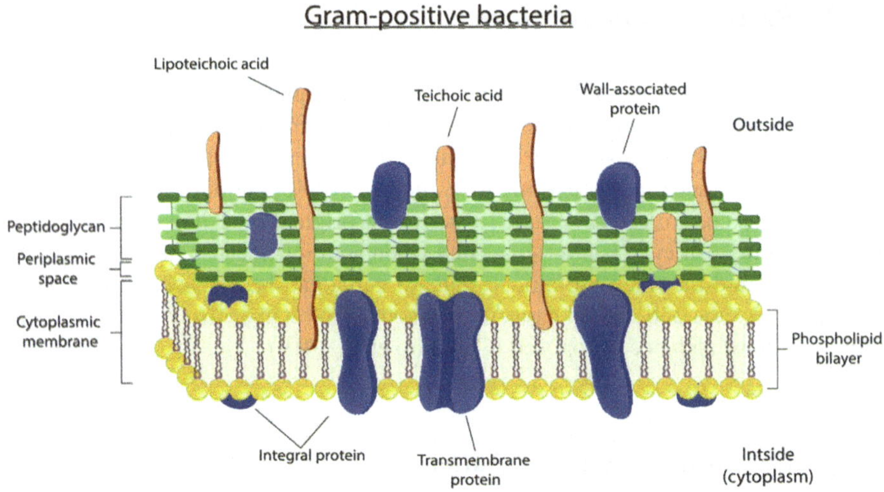

Fig. 1.3 Model of the cell wall structure in Gram-positive bacteria

membrane (phospholipid bilayer) together with its integral proteins. There are thus three essential layers of the Gram-negative cell envelope which provides a formidable barrier for drug permeation (Rojas et al. 2018). The porins, which are the most abundant proteins of the outer membrane in Gram-negative bacteria, allow for passive transport of different compounds across the outer membrane as well as apparently playing an important structural role.

In contrast, Gram-positive bacteria have a simpler structure with an outer peptidoglycan layer, then a periplasmic space before the cytoplasmic membrane formed from a phospholipid bilayer (Fig. 1.3). Lipoteichoic acid and teichoic acid are also important cell wall constituents as are the wall associated proteins and the proteins associated with the cytoplasmic membrane.

While different ways of approaching intracellular Gram-negative entry have been examined, there were no staightforward sets of rules in terms of physicochemical parameters to follow in view of the conflicting property demands required at different points through the cell wall (Silver 2016, 2011). However, some promising predictive rules for key structural requirements in small molecules have now been proposed for accumulation in *Escherichia coli* and these should help significantly with the development of more effective compounds against this and other problematic Gram-negative bacteria (Richter et al. 2017). These rules or guidelines resulted from an assessment of some 180 structurally diverse compounds known to accumulate in *Escherichia coli* and predicted to enter through outer membrane porins (Fig. 1.2), which make up the main outer membrane proteins in Gram-negative bacteria and allow for the entry through diffusion of compounds into the periplasmic space. Some porins contain a binding site for a specific solute molecule which allows for specific crossing of the outer membrane. From Richter et al.'s work, the key features for high accumulation included the presence of non-sterically encumbered amino group functionality, some non-polar structural features, molecular rigidity (few rotatable bonds) and low globularity. It was also shown that by including a primary aminomethylene substituent group in the DNA gyrase inhibitor, deoxynybomycin, it increased it's intracellular accumulation over three fold and antibacterial potency in vitro against a range of laboratory and clinical *Escherichia coli* strains, as well as against strains of *Acinetobacter baumanii, Klebsiella pneumoniae, Enterobacter cloaca*, and against *Pseudomonas aeruginosa.* In a number of cases the potency shown was better than that of the control, ciprofloxacin. Interestingly, strong activity was also seen against two strains of the Gram-positive pathogen *Staphylococcus aureus* (Richter et al. 2017). Further discussion of the new rules for Gram-negative penetration is given in Chap. 3, Sect. 3.1.2.

1.3.4 Other Survival Strategies

Another protective mechanism is associated with responses by bacteria to stress, including assault by antibiotics, which can involve adaptive changes like going into 'dormancy' for the time of exposure to an antibiotic then re-growing when exposure is stopped after antibiotic removal. This is a tolerance strategy which enables bacterial survival, as long as the exposure to the antibiotic is not too prolonged (Fridman et al. 2014). The mechanisms of collective antibiotic tolerance and possible intervention strategies have been elaborated in a good article by Meredith et al. (2015). Further study to determine what sensors are involved in triggering tolerance could potentially be useful in informing the design of other potential multi-targeted ligands which might interfere with any signalling process or processes involved. While not impossible, it will be challenging to design agents to overcome antibiotic tolerance in tackling resistance and it will be important in antibacterial screening methodology to include drug-tolerant bacteria (Stokes et al. 2019).

Further resistance is manifested in slow growing bacteria through the expression of persister cells (Kåhström 2014) and also with the formation of spores which can be hard to counter. Sporolation can complicate the treatment of chronic bacterial infections as is the case with *Clostridium difficile* (Jarrad et al. 2015).

Bacterial persistence is defined as the development of antibiotic-tolerant slow-growing persister cells as a sub-population within the bacterial population resulting in difficulties with responses to antibiotic treatments and a biphasic killing curve (Carvalho et al. 2019). Interestingly, another phenotypic response which seems to be related to, but not the same as, persistence in some ways is that identified as the Eagle Effect in which bacteria can have a higher level of survival when exposed to antibacterial drug concentrations at higher levels than an optimal bactericidal concentration. Under these conditions there is a net decrease in the rate of cell death (Prasetyoputri et al. 2019).

An important and highly problematic further bacterial protective mechanism is that of biofilm formation. This involves a multistep process and the ultimate protection of bacteria on surfaces by a covering layer which generally provides a shield from exposure to an antibacterial agent (Borges et al. 2015). Ways to disperse biofilms and expose the bacteria to antibacterials is an important area of current research.

Another bacterial survival strategy is that of programmed cell death in bacterial communities which results in the death of some bacteria to ensure the survival of the greater population (Peeters and de Jonge 2018; Tanouchi et al. 2013); see also Sect. 5.4 in Chap. 5. While it is an advantageous strategy, it may also open up a potential bacterial vulnerability in which unprogrammed, or premature, cell death might be initiated via small molecule interventions, which could be detrimental. Careful structural design work would be required for such small molecules but it is considered worth pursuing. To inform the design work one might search initially for endogenous molecules in bacteria which sense premature bacterial cell death and which then initiate other actions to locate and nullify the problem. Interference with the actions of any such molecules, if identified, would then need to be factored in to the design process.

1.4 Approaches to Meeting Needs

Although many lines of enquiry aimed at countering human bacterial disease threats are now being pursued (Thayer 2016; Brown and Wright 2016), additional radical ideas and new approaches are required, as well as improvements on current treatments. For example, the disease induced by *Clostridiodes difficile* is a major health care threat in the US (CDC 2019) and elsewhere, and although it is not related to antibiotic resistance acquisition, there is a compelling need for bacterially selective therapeutics for this pathogen (CDC 2019).

Approaches to meeting such needs cover a number of broad strategic categories including the following: theranostics (a combination of therapeutics and diagnostics); rational improvement of existing antibiotics; remedying the liabilities of old antibiotics; and looking for new antibacterials with different structures and preferably different modes of action (Cooper 2018). Approaches in the last category can include screening the chemically diverse global compound libraries for example via the Community for Open Antimicrobial Drug Discovery (CoADD) (Blaskovich 2016) and through looking at new natural product sources, as well as devising and implementing new molecular designs.

A good example of liability repair is the replacement of a secondary hydroxyl group by a fluoro group in the aminoglycoside, neomycin B, an old antibiotic, but one now susceptible to aminoglycoside-modifying enzymes which acetylate amino groups and phosphorylate hydroxyl groups. Such modifications compromise potency through consequent changes in polarity and thus cell penetration. A series of 4′-deoxy-4′-fluoro neomycin analogues were made with improved resistant enzyme inhibition profiles while retaining good antibacterial potency in vitro (Hanessian et al. 2014). An interesting variation on substituent variation to improve potency also involves the neomycin template and attachment of a catalytic diamino alkyl group via a 4′-ether linkage at one of the terminal sugar units. After binding of the neomycin to the bacterial rRNA A-site it is proposed that the warhead is then positioned to catalyse the cleavage of a phosphodiester bond between a guanidine and adenine unit thus impairing the bacterial ribosomes (Smolkin et al. 2019). Encouragingly, this derivative was particularly active against problematic strains of *Pseudomonas aeruginosa* and methicillin-resistant *Staphylococcus aureus* (MRSA).

To try and discover and develop radically new antibacterials (Gadakh and Aerschot 2015; Gilbert 2018; Theuretzbacher 2020b) with new modes of action is another option but this is often a complex and difficult process. The insightful and detailed perspective review by Lewis (2020) highlights this complexity and indicates promising evolving strategies to tease out key elements of the discovery process. From their new approach involving looking for new antibiotics from unculturable bacteria, the Lewis group discovered teixobactin, an antibiotic with a new dual mode of action (Ling et al. 2015).This antibiotic was discovered from the elegant in situ approach to bacterial cultivation introduced by Lewis and Epstein. Teixobactin acts by attacking two critical cell wall components lipid II (Wen et al. 2018) and lipid III, but unfortunately it is not active against Gram-negative pathogens. Teixobactin is discussed further in Chap. 3, Sect. 3.2.2 in the context of new hybrid design.

Mining the so-called microbial dark matter via a number of approaches has also been used by other investigators looking for new antibiotics from new bacteria, including amongst others single–cell genetic sequencing and other techniques to identify uncultured bacteria, as well as the different approach of analysing whole banks of bacterial data for example from the human microbiome (Lok 2015).

Another approach towards meeting the needs arising from bacterial resistance is one based on assessing new combinations of known drugs where drugs can be improved in efficacy through combination with others as separate compounds. An example of such an approach is the development of the synergistic triple combination of the approved β-lactam drugs meropenem (a carbapenem), piperacillin (a penicillin) and tazobactam (a bacterial β-lactamase inhibitor) which is effective in vitro against MRSA, including clinical strains, and against the MRSA strain N315-induced infection in vivo in a mouse model. These drugs have compromised antibacterial efficacy when administered alone (Gonzales et al. 2015). This combination approach, particularly involving three or more drugs, is discussed further in Chap. 2.

A different line of enquiry hinges on developing or re-purposing known drugs used for other indications into use as antibacterials. Such agents can be re-purposed as is or serve as the basis for further structural change to improve potency. For example, the approved anti-rheumatic drug auranofin (Fig. 1.4), a thia glucose-gold derivative, has been shown to exert potent antibacterial activity both in vitro and in vivo against

Fig. 1.4 Structure of auranofin

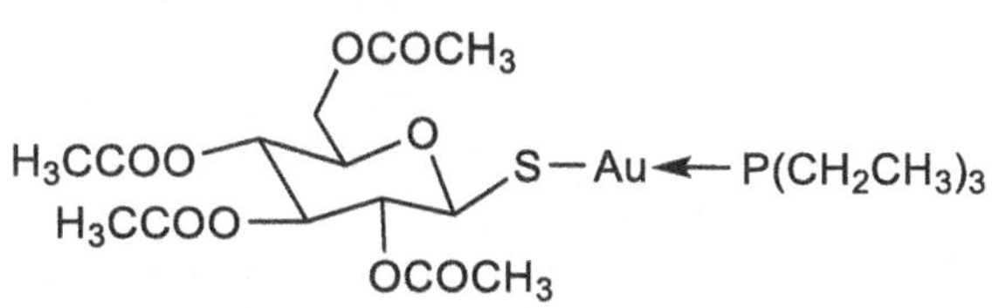

Fig. 1.5 Structure of halicin

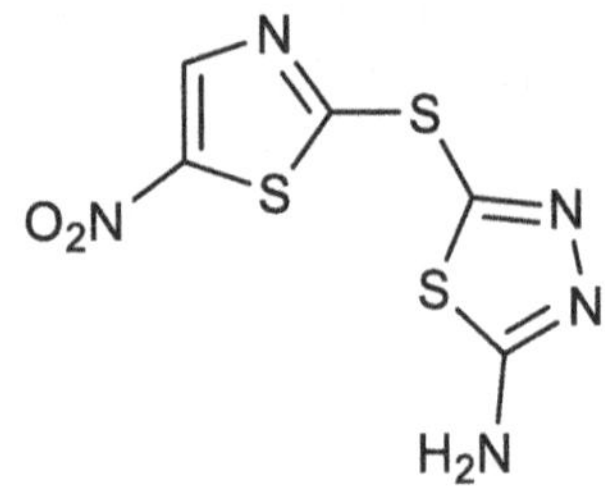

multidrug resistant pathogenic bacteria, and a number of target sites are implicated including in the cell wall, DNA and protein synthesis (Thangamani et al. 2016) (see also Sect. 3.4 in Chap. 3).

A comprehensive recent review by Farha and Brown covers the area of drug re-purposing in the development of new antibacterials and the advantages and disadvantages of this approach are clearly expounded (Farha and Brown 2019). As perhaps to be expected, drug re-purposing is not a perfect solution but it is possible that drugs used in other treatments could reveal new targeting possibilities which could then be considered in further multiple action antibacterial design. However, as is often the case in new drug design, for every plus though there is a minus, and it is the relative magnitudes of each which need to be carefully assessed. A summary of the re-purposing approach in connection with antimicrobial resistance is also given in Kaul et al. (2019), and a review of re-purposed drug possibilities in the therapy of tuberculosis has been detailed by An et al. (2020).

Opportunities are arising in the drug re-purposing space for big data and artificial intelligence (AI) to meet needs by being used to identify new antibacterials amongst compounds prescribed for other conditions. The use of AI in fact was the case when the drug halicin was also discovered to be an antibacterial through an interesting trained deep neural network approach which predicted antibiotic activity in structurally different drug molecules used for other therapeutic purposes (from the ZINC15 database; Drug Repurposing Hub). Halicin (Fig. 1.5) is the name given to SU-3327, a selective kinase (JNK) inhibitor and which was investigated for the treatment of diabetes. In further testing, halicin was shown to have broad-spectrum antibiotic activity in mice. Halicin is believed to act via selective dissipation of the differential pH component of the proton motive force across the bacterial cell membrane. It is also suggested that halicin may bind Fe^{3+} prior to association with the cell membrane (Stokes et al. 2020). If analogues could be developed which significantly reduced the availability of Fe^{3+} while retaining the selective cell membrane interaction they could be very potent new antibacterials. This general approach which can be incorporated into hybrid design is discussed further in Chap. 3.

Aside from random screening of known drugs in combination with others, it is suggested that one might also deliberately assess results (retrospective as well as prospective assessment) when one or more drugs are taken or prescribed at the same time as an antibiotic and look for any potentiating effects on progression of the bacterial disease noted clinically. A previously undetected beneficial drug interaction could be useful in the design of new combination treatments for bacterial disease but negative effects would also need to be carefully considered. This approach may also provide opportunities for suggestions of new combinations for multiple activity.

As alternatives to antibacterial agents, vaccines and other biological approaches are also being actively pursued to help address the needs but these are outside the scope of this book for detailed treatment. Such approaches are complementary to new small molecule development. Vaccines provide the trigger for the production of antibacterial 'antibody agents' in situ through the immune system but biological macromolecules are involved. Vaccine approaches have been successful in fighting some bacterially-induced diseases, including for example Menigococcal meningitis and Pneumococcal disease amongst others. A number are also in the pipeline including one from Pfizer which has a prophylactic vaccine PF-06425090 in Phase III development for *C. difficile* infections and results are due later in 2020 (Shen and Cooke 2019).

1.5 Ways to Achieve Multi-action Effects

While much effort in medicinal chemistry in the past has been directed toward the development of therapeutic agents with one main site of binding to then mediate the desired biological effect(s), the possibilities inherent in deliberate or intentional design for more than one action is rightly an increasing focus of research activity, including in the antibacterial space. The emphasis in this book is on compounds with three or more synchronous or near synchronous actions or multi-targeting ability. This deliberate design paradigm, which is also known as a multivalent approach (Long et al. 2008), is, along with multiple drug combinations whose components interact with different biological targets but are normally administered simultaneously or near simultaneously, part of the area referred to in general as polypharmacology. The case for polypharmacology is a convincing one (Hopkins 2012) and the area, including applications in antibacterial drug discovery (Silver 2012; Brötz-Oesterhelt and Brunner 2008), has been well reviewed by a number of authors (Morphy 2012a, b; Bolognesi 2013; Peters 2013).

Polypharmacology continues to develop as a trend in antibacterial discovery and is gaining considerable contemporary traction with good reason. Key aspects of this area are covered in the excellent recent review by Gray and Wenzel (2020) on multi-target approaches to counter drug resistant bacteria. The strategies employed in the deliberate design of multi-target therapeutics are clearly outlined with specific examples in this review, while Zhou and co-authors have also discussed rational design aspects for multi-target directed ligands or 'designed multiple ligands' (Morphy and Rankovic 2005) for a range of potential disease-treatment applications (Zhou et al. 2019). Single molecule hybrids with more than one action are also referred to as multifunctional compounds and these have been well developed for the treatment of multi-factorial diseases. Zhou et al. (2019) also briefly discuss, in the infectious disease area, dual inhibitors of the integrase and reverse transcriptase enzymes required for human immunodeficiency virus (HIV) replication, as well as details of a hybrid compound based on the anti-malarial agents quinine and artemisinin which displayed potent in vitro activity against against two drug resistant strains of *Plasmodium falciparum*.

A useful overview of multi-target drugs against *Mycobacterium tuberculosum* is contained in the review by Viana et al. (2018) and similarly in the earlier treatment by Scotti et al. (2016) on multi-target drugs for the treatment of AIDS and tuberculosis.

A key aim of this book is to emphasise ideas and possibilities for the design of multi-target molecules for triple or greater antibacterial activities. As an alternative to multi-action combinations of two or more compounds, one can also envisage multiple activities for a single non-cleavable hybrid agent. These single molecules may be hybrids or agents which remain intact after administration and prior to interacting with biological targets associated with bacteria, staying intact at each target site. Alternatively, prodrugs could be designed which are selective for bacteria and are cleaved in or near the bacteria to give two, or subsequently more, products with different synchronous or near-synchonous actions through bacterial target interactions (Bremner 2017). Such molecules could also be components of combinations which may include single action antibacterials as well.

The design of such multi-targeted compounds can be problematic and sometimes there is a contradictory combination of parameters required for single molecule design and one may need to consider quite new and intricate types of structures. However, in this as in other design areas, much can be gained by working meticulously on a small canvass while letting the imagination flourish—think of the powerful beauty and effect achieved by Dutch artist Johannes Vermeer in the masterpiece 'Girl with a Pearl Earring'. The painting is only quite small but is painted to such radiant effect and attention to detail. The important thing is to think widely and curiously outside the square before filling in the essential details. To start with partial ideas or 'stubs' and then build on these. Many will fail to develop, but a few might, which is the important thing. With bacteria we are dealing with very small but intricate biological systems and detailed thinking is required to multiply block or attenuate these systems.

Extensive potential exists for the further design and development of such hybrids or prodrugs and detailed exploration of these ideas with both actual and potential examples is a key unifying theme of this book.

In this book, the terms action and activity refer to the effects on the bacterium exposed to the antibacterial or non-direct antibacterial (or adjuvant) as a result of interaction with the bacterial molecular target or target site(s). This process is referred to as targeting. The site might involve biomacromolecular systems or smaller endogenous molecules such as quorum sensing signalling agents. One should also note that there could also be more than one binding site per target molecule. Lange et al. (2007) have discussed targeting and presented useful data on different bacterial targets, including binding site numbers. Deliberately designed single molecules acting on multiple targets (multiple target compounds) or displaying two or more pharmacological activities in a 'multifactorial disease' (multifunctional compounds) have been discussed by Bansal and Silakari (Bansal and Silakari 2014). Included in this review, for example, is a discussion of berberine-derived hybrid multifunctional compounds as potential drugs for the treatment of Alzheimer's disease through multiple activities including suppression of Aβ protein aggregation, antioxidant activity, and acetylcholinesterase and butyrylcholinesterase inhibition. A tacrine-based trihybrid multifunctional derivative has also been described which incorporates a terminal ONO_2 moiety as a nitric oxide (NO) donor; the NO released in vivo could then also have further activities (Bansal and Silakari 2014). The general principles elaborated in the design of these potential drugs for Alzheimer's disease could potentially be utilised in the design of triple or higher action drugs for the treatment of bacterial diseases.

To exemplify this more specifically for triple action agents, this could mean theoretically interacting with: (i) three separate sites (A′, B′, C′) on the one biomolecule (e.g. a protein) or (ii) two different biomolecules, one with one interacting site (A′) and the other with two different sites (B′, C′), or (iii) three different target biomolecules each with a different single binding site (A′ or B′ or C′), plus permutations of the combination A′, B′, C′. A total of six permutations are possible in type (i) assuming no asymmetric elements are involved. With just two sites (e.g., B′, C′) on the one biomolecule, then a total of four permutations are possible. The resultant activities may be directly antibacterial or have indirect effects increasing design possibilities. In the case of three possible targets one or two of them could be indirect while still maintaining one direct antibacterial interaction or all three could be indirect but ultimately result in bacterial death or stasis. These sites may be intra-cellular, in the cell wall, or near-extracellular.

But, as mentioned, one can also consider combinations of drugs rather than hybrids to achieve multi-targeting objectives. Combinations can involve the administration of directly active antibacterials with or without indirectly acting compounds or adjuvants. The actions of different antibacterials or adjuvants may be synergistic (potentiating) (Bottegoni and Cavalli 2017), have no net increased or decreased effect (additive), or be antagonistic. Multi-action combinations are discussed in further detail in Chap. 2 of this book and their significance in informing the design of potential multi-action hybrids is considered in Chap. 3.

One of the proposed key advantages of multiple action single agents (administered as is or as prodrugs) or combinations of single agents is that bacterial resistance is less likely to develop. However it should be noted that development of resistance is still feasible even though less likely. For example it has been shown that resistance to the dual action antibiotic gepotidacin (a triazaacenaphthylene derivative; GlaxoSmith Kline), which selectively inhibits both bacterial DNA gyrase and topoisomerase IV enzymes involved in bacterial replication, can develop in *Klebsiella pneumoniae* via stepping mutations i.e. a combination of two specific mutations (but not both at the same time which would be difficult). Also the resistant mutant *Klebsiella pneumoniae* was still as virulent as the wild type susceptible strain in a mouse model pathogenicity test (Szili et al. 2019). Also in dual combination treatments resistance can develop in one of the two components as seen with the dual ceftriaxone and azithromycin treatment trial of gonorrhoea in Australia in 2014, which indicated susceptibility to ceftriaxone was retained, but saw a worrying rise in the resistance to azithromycin of *Neisseria gonorrhoea* (Autralian Commission on Safety and Quality in Health

Care 2018). Also resistance to the last-resort antibiotic ceftriaxone (Vincent et al. 2018) has been reported in which the bacterium can maintain a good growth rate via ‘compensatory’ mutations to counter the growth-slowing effect of the resistance mutations. This is a real concern and highlights again the need for new antibacterials for this bacterium.

1.6 Bacterial Over Host Selectivity

Selective toxicity for bacterial cells over host mammalian cells in vivo is a key issue for both combination and single molecule approaches with the antibacterials being launched into a sea of other small molecules, together with many macromolecules both soluble and membrane bound. Important factors include the mode of administration followed by the selective targeting of bacterial cells (intracellular) or contiguous areas (extra-cellular), as well as considering interactions with bacterially specific targets for the mediation of the antibacterial activity. Intracellular targeting can be achieved by passive targeting and passive uptake, passive targeting and selective uptake (compared with host cells), or by intentional or deliberate targeting and then uptake. Following this the molecule may then be released with intact antibacterial activity (Single molecule non-cleavable multiply active antibacterials; Chap. 3) or be transformed or cleaved (Prodrugs for multiply active antibacterials; Chap. 4). Similar considerations apply with extra-cellular agents and combinations (Chap. 2) with either or both intra- and extra-cellular modalities. In Chap. 5 future perspectives are considered to conclude the book.

References

Almeida A, Mitchell AL, Boland M et al. (2019) A new genomic blueprint of the human gut microbiota. Nature 568 (7753):499–504

An Q, Li C, Chen Y et al. (2020) Repurposed drug candidates for antituberculosis therapy. Eur J Med Chem 192: Article 112175

Australian Commission on Safety and Quality in Health Care (2018) Antibiotic resistant gonorrhoea drives bulk of rise in alerts. https://www.safetyandquality.gov.au/media_releases/antibiotic-resistant-gonorrhoea-drives-bulk-of-rise-in-alerts. Accessed 29 Nov 2020

Bansal Y, Silakari O (2014) Multifunctional compounds: smart molecules for multifactorial diseases. Eur J Med Chem 76:31–42

Blair JMA (2018) A climate for antibiotic resistance. Nature Clim Change 8:458–461

Blair JMA, Webber MA, Baylay AJ et al. (2015) Molecular mechanisms of antibiotic resistance. Nat Rev Microbiol 13:42–51

Blaskovich M (2016) Keeping up with the microbes. Chem Aust April:16–19

Bolourian A, Mojtahedi Z (2018) *Streptomyces*, shared microbiome member of soil and gut, as “old friends” against colon cancer. FEMS Microbiol Ecol 94(8): fiy120, doi:https://doi.org/10.1093/femsec/fiy120

Bolognesi M L (2013) Polypharmacology in a single drug: Multitarget drugs. Curr Med Chem 20: 1639–1645.

Borges A, Saavedra MJ, Simões M (2015) Insights on antimicrobial resistance, biofilms and the use of phytochemicals as new antimicrobial agents. Curr Med Chem 22 (21):2590–2614
Bottegoni G, Cavalli A (2017) Computational methods in multitarget drug discovery. In: Decker M (ed) Design of hybrid molecules for drug development. Elsevier, Amsterdam, pp 239–258
Breijyeh Z, Jubeh B, Karaman R (2020) Resistance of Gram-negative bacteria to current antibacterial agents and approaches to resolve it. Molecules 25 (6):1340–1362
Bremner JB (2017) Molecular design of potential triple-action antibacterial agents and related pro-drugs. SWU Sci J 33 (1):1–19
Brötz-Oesterhelt H, Brunner NA (2008) How many modes of action should an antibiotic have? Curr Opin Pharmacol 8 (5):564–573
Brown ED, Wright GD (2016) Antibacterial drug discovery in the resistance era. Nature 529:336–343
Butler MS, Blaskovich MA, Cooper MA (2017) Antibiotics in the clinical pipeline at the end of 2015. J Antibiot (Tokyo) 70 (1):3–24
Butler MS, Paterson DL (2020) Antibiotics in the clinical pipeline in October 2019. J Antibiot (Tokyo) 73 (6):329–364
Carvalho G, Forestier C, Mathias J-D (2019) Antibiotic resilience: a necessary concept to complement antibiotic resistance? Proc R Soc B 286:20192408
CDC (2019) Antibiotic threats in the United States, 2019. Centers for Disease Control and Prevention, Atlanta. https://www.cdc.gov/drugresistance/biggest-threats.html. Accessed 2 Mar 2020
Chen L, Todd R, Kiehlbauch J et al. (2017) *Notes from the field:* Pan-resistant New Delhi metallo-beta-lactamase-producing *Klebsiella pneumoniae*-Washoe County, Nevada, 2016. MMWR Morb Mortal Wkly Rep 66: 33
Cooper MA (2018) Approaches to combat drug-resistant pathogens. Paper presented at the RACI Organic Division National Conference Organic 18, University of Western Australia, Perth, 2–6 Dec 2018
Dixit R, Suseela MR (2013) Cyanobacteria: Potential candidates for drug discovery. Antonie van Leeuwenhoek 103 (5):947–961
Farha MA, Brown ED (2019) Drug repurposing for antimicrobial discovery. Nat Microbiol 4 (4):565–577
Fridman O, Goldberg A, Ronin I et al. (2014) Optimization of lag time underlies antibiotic tolerance in evolved bacterial populations. Nat 513:418–421
Gadakh B, Van Aerschot A (2015) Renaissance in antibiotic discovery: some novel approaches for finding drugs to treat bad bugs. Curr Med Chem 22 (18):2140–2158
Gajdács M (2019) The concept of an ideal antibiotic: Implications for drug design. Molecules 24:892
Garcia-Gutierrez E, Mayer MJ, Cotter PD et al. (2019) Gut microbiota as a source of novel antimicrobials. Gut Microbes 10 (1):1–21
Gilbert N (2018) One step ahead. Old drugs and new tricks are helping to restock the antibacterial armoury. Nature 555: 55–57
Gonzales PR, Pesesky MW, Bouley R et al. (2015) Synergistic, collaterally sensitive β-lactam combinations suppress resistance in MRSA. Nat Chem Biol 11 (11):855–861
Gorityala BK, Guchhait G, Goswami S et al. (2016) Hybrid antibiotic overcomes resistance in *P. aeruginosa* by enhancing outer membrane penetration and reducing efflux. J Med Chem 59 (18):8441–8455
Gray DA, Wenzel M (2020) Multitarget approaches against multiresistant superbugs. ACS Infect Dis 6:1346–1365
Hanessian S, Saavedra OM, Vilchis-Reyes MA et al. (2014) Synthesis, broad spectrum antibacterial activity, and X-ray co-crystal structure of the decoding bacterial ribosomal A-site with 4′-deoxy-4′-fluoro neomycin analogs. Chem Sci 5 (12):4621–4632

Herbrík A, Corretto E, Chroňáková A et al. (2020) A Human Lung-Associated *Streptomyces* sp. TR1341 Produces Various Secondary Metabolites Responsible for Virulence, Cytotoxicity and Modulation of Immune Response. Front Microbiol 10 (3028)
Hopkins AL (2012) The case for polypharmacology. In: Peters J-U (ed) Polypharmacology in Drug Discovery. 1st edn. John Wiley & Sons. Inc., Hoboken, New Jersey, pp 1–6
Jarrad AM, Karoli T, Blaskovich MA et al. (2015) *Clostridium difficile* drug pipeline: challenges in discovery and development of new agents. J Med Chem 58 (13):5164–5185
Kåhström CT (2014) Persisters come under fire. Nat Rev Drug Disc 13:19
Kaul G, Shukla M, Dasgupta A et al. (2019) Update on drug-repurposing: is it useful for tackling antimicrobial resistance? Future Microbiol 14:829–831
Kim H, Lantvit D, Hwang CH et al. (2012) Indole alkaloids from two cultured cyanobacteria, *Westiellopsis* sp. and *Fischerella muscicola*. Bioorg Med Chem 20(17):5290–5295
Koulenti D, Song A, Ellingboe A et al. (2019) Infections by multidrug-resistant Gram-negative Bacteria: What's new in our arsenal and what's in the pipeline? Int J Antimicrob Agents 53 (3):211–224
Lange RP, Locher HH, Wyss PC et al. (2007) The targets of currently used antibacterial agents: lessons for drug discovery. Curr Pharm Des 13(30):3140–3154
Lee JYH, Monk IR, Gonçalves da Silva A et al. (2018) Global spread of three multidrug-resistant lineages of *Staphylococcus epidermidis*. Nat Microbiol 3:1175–1185
Lewis K (2020) The science of antibiotic discovery. Cell 181 (1):29–45
Ling LL, Schneider T, Peoples AJ et al. (2015) A new antibiotic kills pathogens without detectable resistance. Nature 517:455–459
Lok C (2015) Mining the microbial dark matter. Nature 522:270–273
Long DD, Aggen JB, Christensen BG et al. (2008) A multivalent approach to drug discovery for novel antibiotics. J Antibiot 61(10):595–602
MacFadden DR, McGough SF, Fisman D et al. (2018) Antibiotic resistance increases with local temperature. Nat Clim Change 8:510–514
Meredith HR, Srimani JK, Lee AJ et al. (2015) Collective antibiotic tolerance: mechanisms, dynamics and intervention. Nat Chem Biol 11 (3):182–188
Mishra SK,Tripathi G, Kishore N et al. (2017) Drug development against tuberculosis: Impact of alkaloids. Eur J Med Chem 137: 504–544
Morphy JR (2012a) Historical strategies for lead generation. In: Morphy JR, Harris CJ (eds) Designing multi-target drugs. The Royal Society of Chemistry, pp 111–129
Morphy JR (2012b) Selective multitargeted drugs. In: Peters JU (ed) Polypharmacology in drug discovery. John Wiley & Sons, Inc., Hoboken, New Jersey, pp 245–262
Morphy JR, Rankovic Z (2005) Designed multiple ligands. An emerging drug discovery paradigm. J Med Chem 48 (21):6523–6543
Munita JM, Arias CA (2016) Mechanisms of antibiotic resistance. Microbiol Spectrum 4(2):VMBF-0016–2015
Peeters SH, de Jonge MI (2018) For the greater good: Programmed cell death in bacterial communities. Microbiol Res 207:161–169
Peters JU (2013) Polypharmacology - foe or friend? J Med Chem 56 (22):8955–8971
Prasetyoputri A, Jarrad AM, Cooper MA et al. (2019) The Eagle effect and antibiotic-induced persistence: Two sides of the same coin? Trends Microbiol 27(4):339–354
Pribis JP, García-Villada L, Zhai Y (2019) Gamblers: An antibiotic-induced evolvable cell subpopulation differentiated by reactive-oxygen-induced general stress response. Mol Cell 74(4):785–800.e7
Raveh A, Carmeli S (2007) Antimicrobial ambiguines from the cyanobacterium *Fischerella* sp. collected in Israel. J Nat Prod 70:196–201
Richter MF, Drown BS, Riley AP et al. (2017) Predictive compound accumulation rules yield a broad-spectrum antibiotic. Nature 545:299–304
Rojas ER, Billings G, Odermatt PD et al. (2018) The outer membrane is an essential load-bearing element in Gram-negative bacteria. Nature 559:617–621

Scotti L, Filho FJ, de Moura RO et al. (2016) Multi-target drugs for neglected diseases. Curr Pharm Des 22 (21):3135–3163
Seipke RF, Kaltenpoth M, Hutchings MI (2012) *Streptomyces* as symbionts: an emerging and widespread theme? FEMS Microbiol Rev 36 (4):862–876
Shen AK, Cooke MT (2019) Infectious disease vaccines. Nat Rev Drug Disc 18:169–170
Silver LL (2011) Challenges of antibacterial discovery. Clin Microbiol Rev 24 (1):71–109
Silver LL (2012) Polypharmacology as an emerging trend in antibacterial discovery. In: Peters J-U (ed) Polypharmacology in Drug Discovery. 1st edn. John Wiley & Sons. Inc., Hoboken, New Jersey, pp 167–202
Silver LL (2016) A Gestalt approach to Gram-negative entry. Bioorg Med Chem 24 (24):6379–6389
Singh SB, Young K, Silver LL (2017) What is an "ideal" antibiotic? Discovery challenges and path forward. Biochem Pharmacol 133:63–73
Smolkin B, Khononov A, Pieńko T et al. (2019) Towards catalytic antibiotics: redesign of aminoglycosides to catalytically disable bacterial ribosomes. Chembiochem 20 (2):247–259
Stinson SC (1996) Drug firms restock antibacterial aesenal. Chem Eng News 74:75–100
Stokes JM, Gutierrez A, Lopatkin AJ et al. (2019) A multiplexable assay for screening antibiotic lethality against drug-tolerant bacteria. Nat Methods 16 (4):303–306
Stokes JM, Yang K, Swanson K et al. (2020) A deep learning approach to antibiotic discovery. Cell 180 (4):688–702.e613
Szili P, Draskovits G, Révész T et al. (2019) Rapid evolution of reduced susceptibility against a balanced dual-targeting antibiotic through stepping-stone mutations. Antimicrob Agents Chemother 63 (9):e00207–00219
Tacconelli E, Carrara E, Savoldi A et al. (2018) Discovery, research, and development of new antibiotics: the WHO priority list of antibiotic-resistant bacteria and tuberculosis. Lancet Infect Dis 18 (3):318–327
Tanouchi Y, Lee AJ, Meredith H et al. (2013) Programmed cell death in bacteria and implications for antibiotic therapy. Trends Microbiol 21 (6):265–270
Thangamani S, Mohammad H, Abushahba MF et al. (2016) Antibacterial activity and mechanism of action of auranofin against multi-drug resistant bacterial pathogens. Sci Rep 6:22571
Thayer AM (2016) Antibiotics: Will the bugs always win? Chem Eng News 94 (35):36–43
The Pew Charitable Trusts (2016) A scientific roadmap for antibiotic discovery. https://www.pewtrusts.org/en/research-and-analysis/reports/2016/05/a-scientific-roadmap-for-antibiotic-discovery. Accessed 24 Jun 2016
The Pew Charitable Trusts (2020) Antibiotics currently in global clinical development. https://www.pewtrusts.org/en/research-and-analysis/data-visualizations/2014/antibiotics-currently-in-clinical-development. Accessed 27 Nov 2020
The Review on Antimicrobial Resistance (2016) O'Neill J (2016) Tackling drug-resistant infections globally: Final report and recommendations. (Chair: O'Neill, J) May 2016: 1–84
Theuretzbacher U, Gottwalt S, Beyer P et al. (2019) Analysis of the clinical antibacterial and antituberculosis pipeline. Lancet Infect Dis 19: e40–50
Theuretzbacher U, Bush K, Harbarth S et al. (2020a) Critical analysis of antibacterial agents in clinical development. Nat Rev Microbiol 18 (5):286–298
Theuretzbacher U, Outterson K, Engel A et al. (2020b) The global preclinical antibacterial pipeline. Nat Rev Microbiol 18 (5):275–285
Viana JdO, Ishiki HM, Scotti MT et al. (2018) Multi-target antitubercular drugs. Curr Top Med Chem 18:750–758
Vincent LR, Kerr SR, Tan Y et al. (2018) In vivo-selected compensatory mutations restore the fitness cost of mosaic penA alleles that confer ceftriaxone resistance in *Neisseria gonorrhoeae*. mBio 9(2):e01905–17
Walsh F (ed) (2015) The multiple roles of antibiotics and antibiotic resistance in nature. FrontiersMedia, SA

Watkins WJ, Plattner JJ (2015) Anti-infective drugs through medicinal chemistry: A 50-year retrospective. In: Desai MC (Ed-in-Chief) 2015 medicinal chemistry reviews, vol 50. Medicinal Chemistry Division of the American Chemical Society, Washington, pp 241–281

Watson T (2019) The trickster microbes shaking up the tree of life. Nature 569:322–324

Wen P-C, Vanegas JM, Rempe SB et al. (2018) Probing key elements of teixobactin-lipid II interactions in membranes. Chem Sci 9:6997–7008

Willyard C (2017) Drug-resistant bacteria ranked. Nature 543:15

World Health Organisation (2017a) Antibacterial agents in clinical development: an analysis of the antibacterial clinical development pipeline, including tuberculosis. (Tacconelli E, Magrini N, Chairs) WHO, Geneva

World Health Organisation (2017b) Global priority list of antibiotic-resistant bacteria to guide research, discovery, and development of new antibiotics. Antibacterial agents in clinical development: an analysis of the antibacterial clinical development pipeline, including tuberculosis. WHO/EMP/IAU/2017.11 Licence: CC BY-NC-SA 3.0 IGO. WHO, Geneva

World Health Organisation (2018) Global tuberculosis report 2018. Licence: CC BY-NC-SA 3.0 IGO

World Health Organisation (2019a) Global tuberculosis report 2019. Licence: CC BY-NC-SA3.0 IGO)

World Health Organisation (2019b) Antibacterial agents in clinical development: an analysis of the antibacterial clinical development pipeline. Licence: CC BY-NC-SA 3.0 IGO. WHO, Geneva

World Health Organisation (2019c) Antibacterial agents in pre-clinical development: an open access database. WHO/EMP/IAU/2019.12 Licence: CC BY-NC-SA 3.0 IGO. WHO, Geneva

Zhou J, Jiang X, He S et al. (2019) Rational design of multitarget-directed ligands: strategies and emerging paradigms. J Med Chem 62 (20):8881–8914

Chapter 2
Antibacterial Combinations

"There is strength in numbers, but organizing those numbers is one of the great challenges."
—John C. Mather, US Physics Nobel Laureate.

Abstract There has been much activity in the area of multiple drug combinations in treating various diseases including bacterially mediated ones. Work in the field is continuing and in this chapter design principles for combinations of separate drugs aimed at more than two types of antibacterial activity or associated activity which contributes to increased potency or overcoming antimicrobial resistance strategies are covered. This chapter concisely reviews previous work over the past decade or so and details current work on combinations of drugs with multiple activities through specific interactions with one or more biological target molecules.

2.1 Introduction

Combinations with multiple actions offer a range of advantages particularly when potential difficulties with differing target sites and pharmacokinetic profiles can be overcome. As Lewis has pointed out in his perspectives article on the scientific basis for antibiotic discovery (Lewis 2020), a vital advantage of combinations centres

J. Bremner, *Multiple Action-Based Design Approaches to Antibacterials*,
https://doi.org/10.1007/978-981-16-0999-2_2

on the potential ability of their component compounds to access a range of targets simultaneously or near simultaneously in contrast to a compound interacting just with a single target, rendering such single compounds more susceptible to resistance development. As well as this, the recognition of effective drug combinations can provide a powerful starting point for single molecule hybrid or prodrug design with potential benefits over the combinations.

Synergistic, additive or antagonistic interactions may occur with combinations. These interactions can be complex and a thorough analysis with respect to drug combinations in general and network perspectives is given in the review by Jia et al. (2009) and in a perspectives article by Chou (2010), who highlights the issues with combination studies.

It is important to note that combinations are not always beneficial and antagonism has been noted between bacteriostatic and bactericidal agents in pairwise drug combinations using 21 different antibiotics at sub-inhibitory concentrations in vitro (Ocampo et al. 2014) and it is not inconceivable that antagonism may be possible with triple and greater combinations. Paradoxically, antagonism may be better though at slowing resistance development. Although it is generally the case that development of resistance may be reduced with combinations it is not always the case. It has been shown that not all synergistic combinations show clinical benefit and in a number of instances do not always slow resistance development. In a later interesting study on species-specific drug combinations, Brochado et al. (2018) looked at nearly 3000 pair-wise combinations of antibiotics, human-targeted drugs and food additives and their effects on three Gram-negative pathogens, namely *Escherichia coli*, *Salmonella enterica* serovar Typhimurium and *Pseudomonas aeruginosa* (a total of six strains from these three pathogens). More than 70% of the identified drug-drug interactions were species specific and some 20% showed strain specificity. Also, significantly, antagonistic interactions were observed more frequently than synergistic ones with the former appearing almost always between drugs targeting different cellular processes in contrast to the latter which were manifested more when the same cellular process was being targeted. This clearly has implications for the design of new multi-targeting combinations.

While the main focus of this book is on ways to achieve more than two actions, looking at dual action drug combinations to start with can provide a useful platform from which to extend to multiply active combinations or the design of new multiply active hybrid molecules or related prodrugs.

In this chapter the classification of the materials is primarily based on the number of separate components in the combinations with sub-sections based on the number of resulting actions or interactions with the bacterial targets or the way in which components might be activated. The actions themselves may be directly antibacterial or they may potentiate this activity through other actions i.e. the component is viewed as indirectly antibacterial. Worthington and Melander (2013) have noted three activity-type categories in antibiotic combinations based on targets and pathways and they also discuss antibiotic with adjuvant combinations. These areas are also covered in the subsequent sections.

2.1.1 *Dual Combinations Resulting in Two Actions*

The area of dual combinations for antibacterial therapy has been, and continues to be, explored extensively and has been reviewed in detail most recently by Domalaon et al. (2018) which includes, amongst other topics, a treatment of dual combinations resulting in two actions. Another excellent recent review in this combination space is that by Tyers and Wright (2019) which also has a good treatment of recent hybrids and the theory associated with synergy plus associated terms. They classify synergistic antibiotic combinations (dual) in three categories: congruous, syncretic and coalistic. With a congruous pair the components target two distinct essential targets, while the syncretic case covers one component targeting an essential bacterial process and the other non-directly antibacterial component interacting with what is designated as a non-essential target or resistance element. In the last category neither component would be antibacterial but could interact with target proteins which correspond to synthetic lethal genetic interaction pairs resulting in specific chemical lethality. Such combinations could result in very narrow-spectrum effects and further developments in this area are likely to emerge in the future. For this section of the book, discussion is concentrated on pairwise combinations in the first two categories. Significantly Tyres and Wright also argue for multi-targeting approaches in new antibacterial designs.

The simplest combination matrix involves the combined administration of two separate drugs. From an in vivo perspective, oral administration could involve two separate tablets or a fixed dose combination in a single tablet (Prati et al. 2014). This combination can be reduced to an A + B notation for drug A and drug B. With this dual combination the two-sub-group classifications would then include in the first sub-group (Sect. 2.1.1.1) the situation where one of the drugs (A) has a direct antibacterial action through a single interaction with a single bacterial target, while the second drug (B) would have a separate potentiating single interaction with another target but not being directly antibacterial. In the second sub-group (Sect. 2.1.1.2) both component (A) and component (B) might be directly antibacterial through a single action each at two different targets or at different points on the same target.

While each drug in dual combinations often has one interaction site each, possibilities exist to extend the drug interaction site to more than one for either one or both components resulting in a triple or higher order interaction spectrum as discussed in the subsequent sub-sections of 2.1.2.

2.1.1.1 A (Antibacterial) and B (Non-direct Antibacterial)

In this sub-group, one component would be directly antibacterial and the other not but potentiating or enhancing the activity of the other component. The latter compound may act within or on the bacterium or outside it, as with quorum sensing antagonists or inhibitors or other anti-virulence activities. The rationale in this particular scenario is that by inhibiting quorum sensing, bacterial pathogenicity will be compromised

Fig. 2.1 Structure of the virulence attenuator MAC-545496

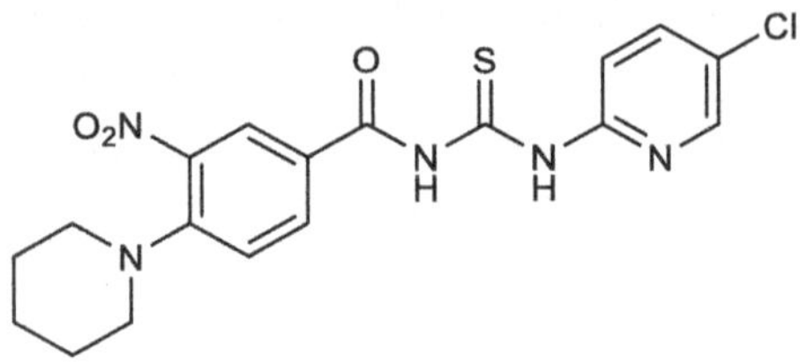

and bacteria may be more susceptible to the action or actions of a direct antibacterial agent. Other potentiating actions include inhibiting enzymes which degrade the antibacterial rendering it inactive or inhibiting bacterial efflux pumps which can reduce substrate antibacterials to sub-MIC intracellular concentrations.

Extension of the interactions to other antivirulence targets by one of the molecular components is also an active research area but this would then mean expansion to a triple combination to enable inclusion in the combination of a directly antibacterial agent as well. Virulence attenuation does not of itself kill the bacterium but can increase the efficacy of antibacterials (Dickey et al. 2017). Other potential advantages of this approach include reducing selective pressure which may reduce resistance development and also having less impact on the microbiome of the host. Antivirulence approaches have been extensively reviewed by Dickey et al. (2017) and by Calvert et al. (2018). An earlier but useful review on combinations is that by Clatworthy et al. (2007) in which approaches to targeting virulence are highlighted.

Illustrative of the potentiating effects of compounds with antivirulence activity is the intriguing small molecule MAC-545496 (Fig. 2.1) which potently reverses β-lactam resistance in *Staphylococcus aureus* (MRSA). This *N*-acylthiourea derivative does not inhibit the growth of the bacterium but does attenuate it's virulence in an infection model based on *Galleria mellonella* (greater wax moth) larvae. In separate assays MAC-545496 was shown to inhibit the survival of MRSA in macrophages and also to lessen the ability of this bacterium to form biofilms. The mechanism of action centres on potent inhibition of GraR (Glycopeptide-resistance-associated protein R), a regulatory protein that is responsive to cell membrane stress and is a significant virulence factor and mediator of antibacterial resistance (El-Halfawy et al. 2020; see also the useful comment by York 2020 on this work).

Other examples of antivirulence agents include the small molecule inhibitors of AgrA (Accessory gene regulator A; Gomes-Fernandes et al. 2017) a protein which is a regulator of gene transcription and the production of secreted virulence factor. Such small molecule inhibitors include the substituted resorcinol-based ketones F12 and F19 (Fig. 2.2), which also potentiate the action of some antibiotics in Gram-positive pathogenic bacteria (Greenberg et al. 2018). One of these antibiotics was the fluoroquinolone sparfloxacin (Fig. 2.3) which interacts with two related enzymes DNA gyrase and topoisomerase IV involved in DNA synthesis and replication (Pham et al. 2019) and thus this mixture constitutes a three action dual combination (Greenberg et al. (2018). Compounds F12 and F19 have been shown to obviate the binding of the

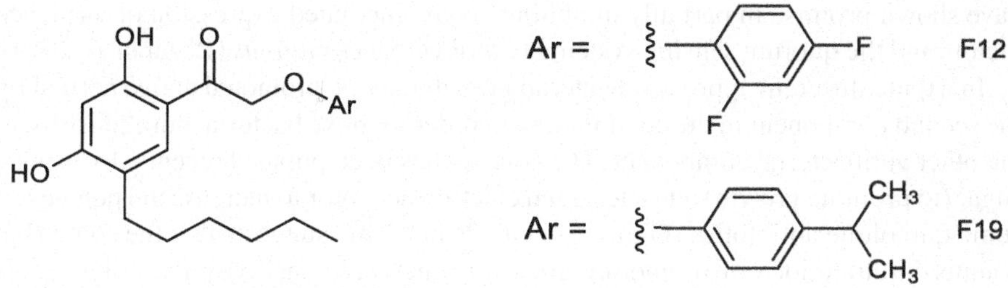

Fig. 2.2 Structures of the AgrA inhibitors F12 and F19

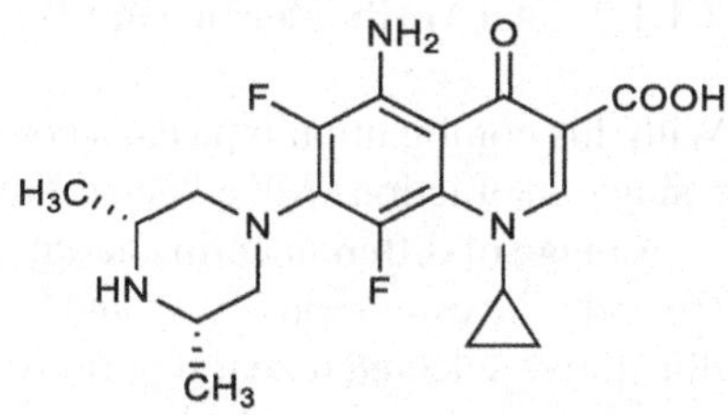

Fig. 2.3 Structure of the fluoroquinolone sparfloxacin

transcription factor AgrA to its promoter in *Staphylococcus aureus*. Further discussion of dual combinations with potentially three-action outcomes is considered in Sect. 2.1.2.

Inhibition of agr quorum sensing in *Staphylococcus aureus* by targeting AgrA with a small molecule, savirin (Fig. 2.4), has also been observed (Sully et al. 2014). Savirin blocks the binding of AgrA to its promoter sites with negative impacts on virulence gene upregulation. Promotion of host defence with little effect on resistance was also seen and this was ascribed in part to increased killing of Agr+ but not ΔAgr in macrophages and to low pH. These results suggest potential for further multiply active combinations. Also drug like fragments, for example 9*H*-xanthene-9-carboxylic acid,

Fig. 2.4 Structure of savirin

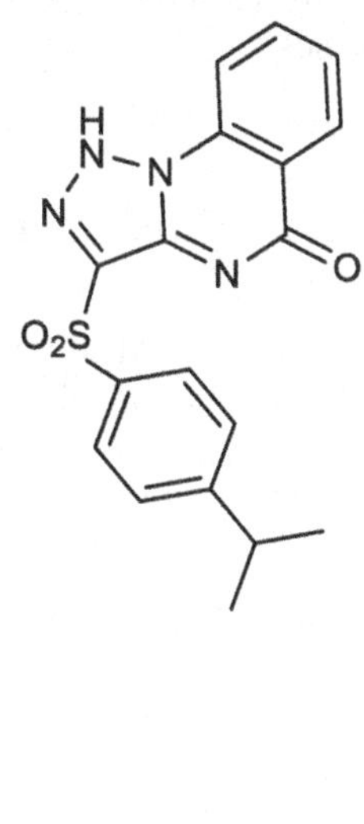

have shown promise in partially inhibiting AgrA- mediated expression of virulence factors and the quorum sensing system in *Staphylococcus aureus* (Bezar et al. 2019).

In a quite different approach, bacterial growth may be promoted above normal by the second component in the combination in order to make bacteria more sensitive to the other antibacterial component. The concept has been pursued recently by using a sugar (to promote growth) and a terminal electron acceptor to increase the potency of some quinolone antibiotics (Gutierrez et al. 2017). This suggests that the somewhat counter-intuitive idea of promoting growth to increase sensitivity to antibacterials has potential to be more widely applied.

2.1.1.2 A (Antibacterial) and B (Antibacterial)

With this combination type each component would be directly antibacterial through a single interaction with a bacterial target in each case.

A range of different antibacterials in such dual combinations have been studied or are under investigation (Tyers and Wright 2019; Domalaon et al. 2018) and only one illustrative selected recent report is discussed here. For example, one line of investigation has been to try and develop trimethoprim analogues to overcome dihydrofolate reductase (DHFR) resistance which compromises the activity of this antibacterial. Trimethoprim has been used extensively clinically in combination with sulfonamides which inhibit dihydropteroate synthetase (DHPS) in the folate pathway. A number of trimethoprim analogues incorporating one or two imidazo-ring fusions to the key pyrimidine core of trimethoprim were prepared. These derivatives incorporated one or both amino substituent group nitrogens of the diaminopyrimidine unit in trimethoprim, and one derivative, a mono imidazo ring-fused analogue, showed promising synergistic in vitro potency in combination with the dihydropteroate synthetase inhibitor sulfamethoxazole in *Staphylococcus aureus* and *Escherichia coli*, although not quite as good as trimethoprim itself. These derivatives could be promising DHFR inhibitor lead compounds (Pedrola et al. 2019). Limitations are apparent though and these new compounds were not active against *Pseudomonas aeruginosa* either alone or in combination with sulfamethoxazole, possibly due to permeability and/or efflux issues.

2.1.2 Dual Combinations Resulting in Three or More Actions

With these dual combinations and three or more target interactions, the classification of combination sub-types becomes quite complex. However, for simplicity one can designate four sub-types using the A, B, C, D notation for each molecular component or part of each component for basic classification purposes as follows:

(i) A + B where one component has two actions and the other one a single action and where at least one of the actions is directly antibacterial for a triple action outcome. This could be extended to a higher order of actions as well.

(ii) (A − B) + C where (A − B) represents a single molecule like a covalently linked dual action hybrid, or a produg which can release dual action compounds in situ, derived from the pharmacophoric units A and B derived from the separate drugs and C is another single or greater action component. Again at least one of the actions from this combination would be directly antibacterial.

(iii) (A − B) + (C − D) where (A − B) and (C − D) each represent a single covalently linked dual action hybrid (or related produg) with dual actions derived from the pharmacophoric units A and B or C and D. At least one of the actions from this combination would be directly antibacterial.

(iv) (A − B − C) + D where (A − B − C) represents a possible triple action single agent (refer to Chap. 3) or prodrug for such an agent (refer to Chap. 4) with the embedded relevant target interaction moieties A, B and C, plus another separate single or higher action compound (D). This is just one representation and there could be a number of variations and extensions on this.

It should be pointed out, however, that some uncertainty can arise in allocating a discussed example to one of these basic classifications if the modes of action are not fully established. Also unrealised interactions may still occur and may influence activity outcomes.

Some specific examples are also included which could potentially express triple or higher activity profiles based on the assessment of other results on alternative components.

2.1.2.1 Dual A Plus B Combination with Three or More Actions

A good example of this A + B group (sub-type i) involves the weak to moderately antibacterial alkaloid berberine in combination with another antibacterial. Berberine (Fig. 2.5), which forms a thematic compound in this book and is discussed in greater detail in Chap. 3 with respect to hybrid design, is thought to exert its activity through a number of mechanisms involving both interacting with the cytoplasmic membrane

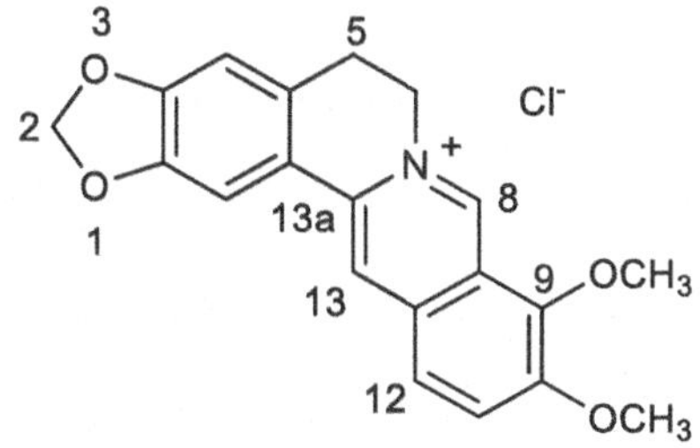

Fig. 2.5 Stucture of berberine, as its chloride salt

and binding to DNA, as well as binding to the crucial FtsZ protein and blocking bacterial cell division. Further references to the mode of antibacterial action of berberine are given in Wang et al. (2016). Antibacterial synergy has been observed with berberine (designated as A) and other antibacterial agents (designated as B) against multi-drug resistant clinical isolates of MRSA (Zuo et al. 2012). Also the combination of berberine with the antibiotic azithromycin was effective against clinical isolates of *Pseudomonas aeruginosa* from the sputum of cystic fibrosis patients and, promisingly, this combination was shown to be active in vivo as well in a mouse model (Shaw and Wuest 2020).

One also needs to differentiate combinations in this sub-type (i) category where one or both compounds have antibacterial activity as well as some other not directly antibacterial activity. For example the dual antibacterial combination of berberine and azithromycin assessed by Shaw and Wuest also inhibited biofilm and virulence factor output through compromising the *Las* and *Rhl* quorum sensing systems (Shaw and Wuest 2020). Ideally the non-directly active antibacterial activity or actions would enhance the direct antibacterial activities and this may be achieved in a number of ways. For example by nullifying resistance mechanisms like those mediated by efflux pumps (Bremner et al. 2007; Brown and Wright 2016) or by aiding antibacterial penetration as in the case of the highly basic drug pentamidine which in Gram-negative pathogenic bacteria disrupts the outer membrane thus allowing entry by the active agent (Stokes et al. 2017).

Blocking efflux pumps can be a very effective potentiating action. There have been intensive investigations of small molecule efflux pump inhibitors and how to achieve such inhibition (Ramaswamy et al. 2017; Haynes et al. 2017; Abdali et al. 2017). Other references to efflux pump inhibitors are also given in Chap. 3 (Sect. 3.3.3). The antibacterial berberine (Fig. 2.5) is a substrate for the NorA efflux pump in *Staphylococcus aureus* but the NorA efflux pump inhibitor INF-55 greatly boosts the antibacterial activity of berberine against this bacterium in vitro when they are used in a combination (Ball et al. 2006).

Another illustrative example in the antibacterial context of this combination sub-type is that of the potent broad spectrum β-lactamase inhibitor Taniborbactam (VNRX-5133; Fig. 2.6a) with the β-lactam antibacterial cefepime (Fig. 2.6b) which binds to penicillin binding proteins (PBPs). The former compound incorporates a 6-membered oxabora heterocycle and inhibits two types of β-lactamases which can

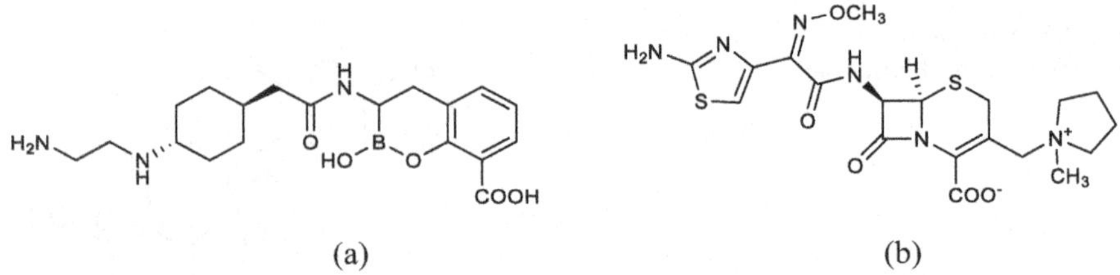

Fig. 2.6 Structure of VNRX-5133 (**a**) and cefepime (**b**)

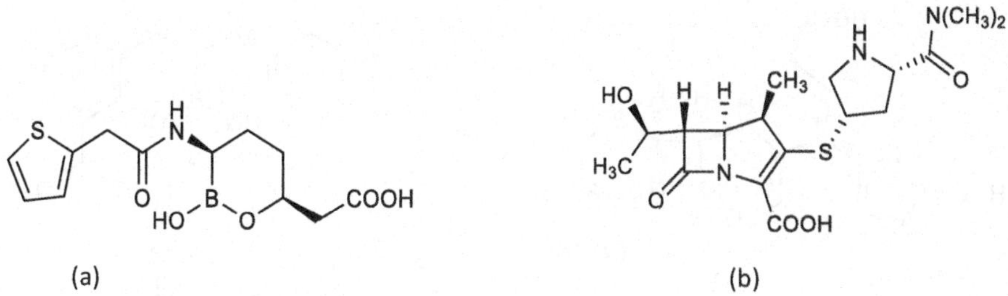

Fig. 2.7 Structures of RPX-7009 (Vaborbactam) (**a**) and meropenem (**b**)

deactivate cefepime. The drug combination is in Phase 3 clinical trials for complicated urinary tract infections (Venatorx Pharmaceuticals 2020; Liu et al. 2020). So in this case the combination is of two compounds with a resultant three actions through three different interaction sites. This combination could thus also provide a potentially good starting point for triple action hybrid design or prodrug design especially if the inhibition of the two β-lactamases was reversible. Both serine- and metallo-β-lactamases can be expressed in the one bacterium as in the case of *Pseudomonas aeruginosa*. In aqueous media, VNRX-5133 is in equilibrium with the ring opened boronic acid-phenol form but studies have shown it is the cyclised structure which interacts with the β-lactamase.

A similar combination treatment is that of the organo-boron derivative RPX-7009 (Fig. 2.7a) together with meropenem (Fig. 2.7b) (Carbavance) (Hernandez et al. 2016). RPX-7009 also inhibits more than one type of β-lactamase while meropenem interacts with a number of PBPs particularly PBP2. The cyclic half boronic acid ester RPX7009 (now known as Vaborbactam), whose discovery was reported by Hecker et al. (2015), is a broad-spectrum β-lactamase inhibitor of particular interest for its inhibition of Class A serine carbapenemases, and following successful clinical outcomes, was approved by the FDA in 2017 for use in combination with meropenem for the treatment of carbapenem-resistant Enterobacteriaceae (Lee et al. 2019). An update review on boronate based lactamase inhibitors and penicillin binding protein (PBP) inhibitors (as inhibitors of cell wall wall biosynthesis), as well as other structural types, summarises useful target site-inhibitor interaction information in this important area (Liu et al. 2019).

Venatorx Pharmaceuticals, the developers of Taniborbactam, which is administered intra-venously with cefepime, have also developed another related oxaborine derivative, the orally bioavailable VNRX-7145 (Fig. 2.8b), which is a prodrug that is converted to the active β-lactamase inhibitor on ester hydrolysis by liver esterases. VNRX-7145 was developed for use in combination with the orally bioavailable cephalosporin derivative, ceftibuten (Fig. 2.8a), for use against extended spectrum and clinically problematic Enterobacteriaceae (Papp-Wallace 2019). The combination has moved to Phase 1 clinical trials (2020) for resistant bacterial urinary tract infections.

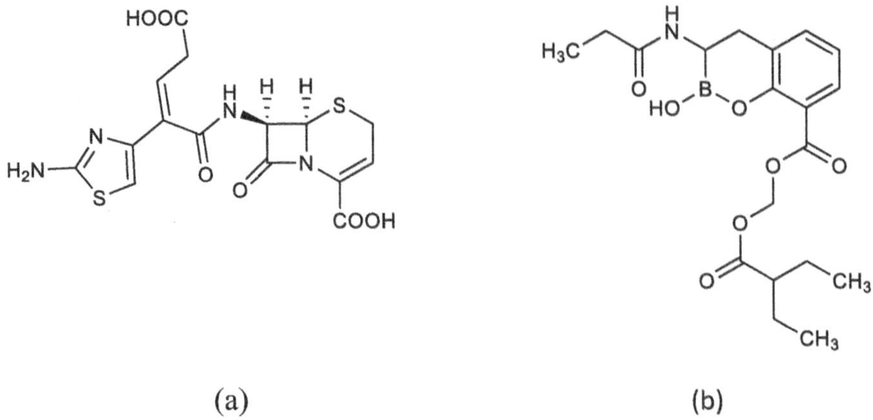

Fig. 2.8 Structures of the cephalosporin derivative ceftibuten (**a**) and the prodrug VNRX- 7145 (**b**)

Fig. 2.9 Structure of the β-lactamase inhibitor QPX7728

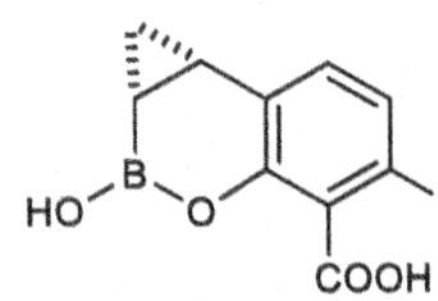

Further development of the cyclic boronic acid derivatives led to the very broad spectrum β-lactamase inhibitor QPX7728 (Fig. 2.9) with potent inhibitory activity against a range of β-lactamases including classes A, B and C, as well as class D (*Acinetobacter*) enzymes coupled with metabolic stability. Potential for intravenous or oral delivery is also indicated (Hecker et al. 2020). An interesting structural feature of QPX7728 is the fused cyclopropane ring on the cyclic boronic acid unit in place of other substituents as present in VNRX-7145 for example.

The incorporation of boron-based components in the synthesis of medicinal agents is a promising area generally and certainly more specifically for the development of new antibacterials. Boron is a versatile element which can form neutral trivalent, and negatively charged tetravalent, structural units each with a different stereochemistry and incorporating bonds particularly to oxygen and nitrogen as well as carbon. In addition boron-centred functional groups can also be effective bioisosteres, for example with the isoelectronic and isosteric BN unit replacing a C=C bond or the B-O unit substituting for C=O (Hernandez et al. 2016).

The American company, Anacor Pharmaceuticals, acquired by Pfizer in 2016, is also developing small molecule therapeutics based on a boron chemistry platform. There has been a considerable revival in synthetic chemistry around boron especially based on what is referred to as frustrated Lewis pairs (FLPs) although high reactivity and stability issues can still be a challenge (Crow 2019). One such therapeutic developed by Anacor/GlaxoSmithKline was the potent bacteriostatic drug Epetraborole

(Fig. 2.10) (AN3365) which is active against Gram-negatives and anaerobic bacteria, including multi-drug resistant strains, through inhibition of leucyl-tRNA synthetase and thus protein synthesis. Inhibition of the synthetase is mediated via the terminal adenine ribose (A 76) in $tRNA^{leu}$ but unfortunately resistance, due to mutations in the leucyl-tRNA synthetase editing domain, has been shown to develop clinically while treating complicated urinary tract infections (O'Dwyer et al. 2015). Clinical trials have now been discontinued on AN3365 (Butler and Paterson 2020). The combination of Epetraborole with another dual action antibacterial would be of interest particularly if it had the potential to reduce the rate of resistance development to the former compound.

Interestingly, boron-based compounds also show other useful activities such as NorA efflux pump inhibition as in the case of compounds based on pyridine boronic acids (Fontaine et al. 2014). As an added feature these derivatives also display some moderate intrinsic antibacterial activity against *Staphylococcus aureus* in vitro. In these compounds the boron was essential for the activity. Such a combination of dual properties suggest the possibility for a new range of combinations as well as single molecule multi-action hybrid development.

A final illustration of the A + B combination having triple or higher order actions is that of antibacterial natural products which have different modes of action but can potentiate the action of the same class of antibiotics. For example both berberine and epicatechin gallates (which inhibit pump efflux modalities and type II fatty acid synthesis in bacteria) can both potentiate β-lactams as referenced in the review by Simões et al. (2009). Berberine has some activity as an MexXY-OprM efflux pump inhibitor in a *Pseudomonas aeruginosa* isolate in a planktonic state, but while the combination of berberine and imipenem (IMP) was synergistic, the activity was still weak against this isolate (Su and Wang 2018). Later it was shown that the synthesized berberine derivative, 13-(2-methylbenzyl)berberine, was a more potent inhibitor of the MexXY pump, and that it acted synergistically with aminoglycoside antibiotics in strains of *Pseudomonas aeruginosa* with higher levels of this pump (Kotani et al. 2019). *In silico* studies point to berberine and the aminoglycoside tobramycin competing for the same site 2 in this pump (Laudadio et al. 2019), but further studies are needed to establish whether the 13-substituted berberine derivative above would bind more strongly to site 2 than seen for berberine.

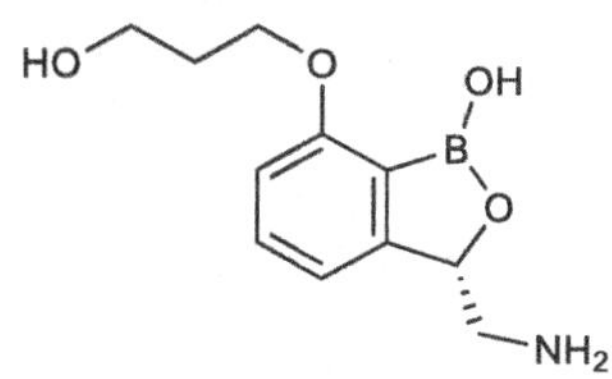

Fig. 2.10 Structure of the leucyl-tRNA synthetase inhibitor AN3365 (Epetraborole)

2.1.2.2 Dual Action Hybrid Agent (A − B) Plus Another Agent (C)

While administering three separate drugs each with a different mode of action can be an effective treatment regime especially where high target doses are required, a potentially viable alternative is to reduce the number of drugs in the combination to two but have one as a hybrid with two mechanisms of action and the other drug with just one action or even more than one. In this way one could have a dual molecular combination of two separate drugs with three, or potentially more, actions. Expressions of this design principle could include perhaps a hybrid peptide (Jindal et al. 2017) plus ceftriaxone, or erythromycin. Other possibilities have also been described including the use of hybrid adjuvants, for instance a tobramycin-ciprofloxacin hybrid compound synergising with mitomycin C as discussed by Domalaon et al. (2019) or of a nebramine-cyclam hybrid conjugate potentiating β-lactam antibiotics in multidrug-resistant *Pseudomonas aeruginosa* (Ammeter et al. 2019). Further studies in this area by the same group now include other nebramine-fluoroquinolone hybrids and an NMP-linked hybrid (NMP, 1-(1-naphthylmethyl)-piperazine, is an efflux pump inhibitor), as described by Yang et al. (2019). Incorporating a hybrid with one other drug in a dual combination could be beneficial in helping to reduce pharmacokinetic issues that might present in vivo with a combination of deconstructed components.

Two-pronged actions at the one site also have potential for new antibacterial design. For example one action could involve specific binding to a biological protein target through non-covalent interactions, but through this interaction favourably dispose a 'warhead' moiety for further covalent bonding involving a neighbouring amino acid residue site with a reactive side-chain like cysteine or serine (Brown and Boström 2018). This strategy has been used to reinforce binding and heighten potency in the oncology therapeutic area with Janus kinase (JAK) inhibitors and should also be transferable to potential bacterial targets as long as selective, but care would be needed. In the antibacterial context, another dual combination possibility would be to incorporate in a single inhibitor molecule bitopic binding to an efflux pump plus an antibiotic with more than one action.

The question arises whether there any examples in Nature of the above dual combination type (A − B) plus (C)? It is not inconceivable that bacteria might produce a dual action antibacterial plus an efflux pump inhibitor as a sophisticated defence against competitive bacteria. This may seem unusual but perhaps the bacterially produced efflux pump inhibitors could be sequestered in some way and then released externally when required together with the antibiotic produced by the bacterium. The production of antibiotics by bacteria is well established as is their use as a weapon against other bacteria in the soil environment; increased antibiotic production has also been noted when there were neighbouring competing bacterial strains (Abrudan et al. 2015). There appears to be no studies reported on possible bacterial co-production of efflux pump modulators and it may be difficult to locate such bacterial producers. Does this occur for example in maintaining a healthy human microbiome with a host plus a bacterial molecular component? Further investigation to try and answer this question would seem warranted. Also of some possible relevance is the fact that

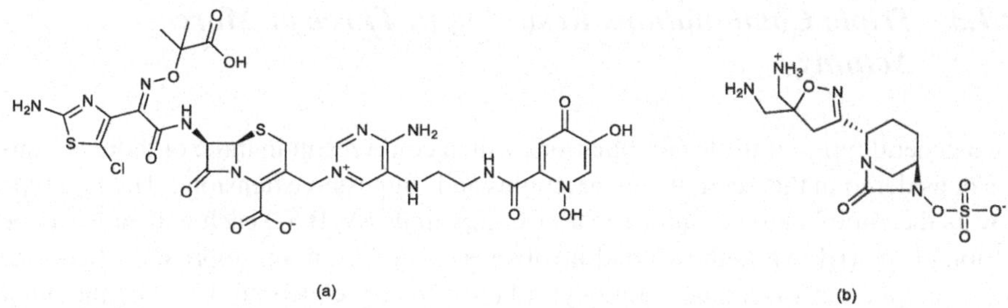

Fig. 2.11 Molecular structures of the hybrid GT-1 (**a**) and of the compound GT-055 (**b**)

plants are known to produce both efflux pump inhibitors and antibacterial agents, presumbably as a defence against plant bacterial pathogens, at least in part.

2.1.2.3 Two Dual Action Hybrids [(A – B) + (C – D)]

In this classification each molecule in brackets represents a single covalently linked dual action hybrid (or related produg) with dual actions derived from the pharmacophoric units A and B and C and D. At least one of the actions from this combination would be directly antibacterial. This basic concept is also suggested in the review by Oldfield and Feng (2014) together with a treatment of approaches to synergy via such combinations.

Various efforts have been made to realise this particular dual combination approach, one example being the combination of the hybrid GT-1 (Fig. 2.11a), a chlorinated analogue related to ceftazidime (PBP binder) which also has attached a siderophore unit, plus the β-lactamase inhibitor and PBP3 binder (in *Escherichia coli* and *Klebsiella pneumoniae*) GT-055 (Fig. 2.11b). This combination can be considered as having the capability for at least four molecular actions (Papp-Wallace 2019; Nguyen et al. 2020). While this combination was in Phase 1 clinical trials earlier the trials have been discontinued.

2.1.2.4 A Triple Action Hybrid (A – B – C) and Another Separate Agent (D)

One of the components in this particular dual combination could be a hybrid construct embodying three recognition sites for target interactions while the other component might exhibit interactions with a further separate target. The design of triple targeting hybrids is discussed in Chap. 3. This is just one representation and there could be a number of variations and extensions on this including the use of produgs as precursors for the active compounds. The general sub-type is raised here though as an option for consideration in future combination designs.

2.1.3 Triple Combinations Resulting in Three or More Actions

Two general types of triple combinations which could result in three or more actions are considered in this section with examples and suggested extensions. The first type covers the combination of three separate compounds (A, B, C) each with at least one action [Type (i)], while the second involves a variation on this approach where one of the components is a dual acting hybrid (A − B) or related prodrug and the other two are separate compounds (C, D) with one or more actions [Type (ii)].

2.1.3.1 Type (I) Combination of Three Separate Components A, B and C

There are a number of examples of the use of combinations of three separate drugs in antibacterial studies and some selections from these studies include the work of Beppler et al. (2016), Hamoud et al. (2014), and Tekin et al. (2016), and the review by Decuyper et al. (2018). Promising outcomes have been reported in terms of effects on bacterial growth including of problematic strains. Triple antibiotic combinations were shown to be effective in vitro against extreme drug resistant *Pseudomonas aeruginosa* and *Acinetobacter baumannii* and that the compounds acted synergistically (Aboulmagd and Alsultan 2014). Synergistic interactions have also been noted with the triple combination of thymol, ethylendiamine tetraacetic acid (EDTA), and vancomycin, including a significant 16-fold enhancement in sensitivity of the Gram-negative pathogenic bacterium *Escherichia coli* (Hamoud et al. 2014). The phenolic monoterpene thymol has antibacterial properties at least in part due to alteration in membrane permeability and in some outflow of intracellular components subsequent to disturbance of the plasma membrane lipid fraction (Trombetta et al. 2005). On the other hand, the glycopeptide antibiotic vancomycin affects cell wall synthesis and is used for the treatment of Gram-positive bacterial infections. EDTA, a chelating agent for Ca^{2+} and Mg^{2+} ions, was included as these ions are important in bacterial cell wall protection (particularly in Gram-negative bacteria) and it was thought that if this protection was compromised by chelation then other antibacterials might be more effective, as was the case.

In other incisive work by Yeh and co-workers (Tekin et al. 2016), the triple drug combination of ciprofloxacin, clindamycin and streptomycin resulted in an antibacterial synergy against *Escherichia coli* (decreasing bacterial fitness-growth rates) in contrast to some other triple antibiotic combinations. Each antibacterial was given at an individual fixed concentration and the synergistic interaction was in accord with both the deviation from additivity measure (DA) and the emergent (E3) interaction algebraic measures. The DA is derived from the two drug framework, while E3 characterises how much of the three-antibacterial interaction is not derived from pairwise interactions. In the case of the three drug combination there would be three such pairwise situations to be considered. While emergent lethal synergy was seen with the triple combination of ciprofloxacin, clindamycin and streptomycin, a significantly suppressive interaction was observed with erythromycin, cefoxitin and tobramycin. What is of further interest here from a design perspective is the possible mechanism of action reasons for synergy versus antagonism since this could help refine the search for new triple combinations and, subsequently, new hybrid or prodrug designs to counter resistance issues. In the cases noted above ciprofloxacin inhibits DNA gyrase and topoisomerase IV, cefoxitin inhibits cell wall synthesis, while the other antibacterials interact mainly either with the 50S (clindamycin; erythromycin) or 30S (streptomycin; tobramycin) ribosomal sub-units. Also of relevance is that drug combinations may be more effective at reducing resistance evolution. Yeh and co-workers have developed (Beppler et al. 2016) a very useful framework to evaluate potentially therapeutically valuable synergies from three-way interactions of separate antibacterials using *Escherichia coli*.

While three (and four) drug combination effects are said to arise from the accumulation of pairwise interactions (Wood et al. 2012), this needs to be reassessed in the light of the work by Yeh's group, which indicates their re-scaled equations for three component combinations could be extended to the quantitative analysis of systems with four components (or higher) and these may give different conclusions. An emergent four way interaction measure has been derived (Beppler et al. 2016). Other foreshadowed developments would be to look at the effect of antibacterial component concentration gradients and timing of addition of each drug on bacterial growth. The concentration gradient studies could conceivably be extended to quantitatively analyse bacterial growth effects in three component mixtures in which one component may be a drug efflux inhibitor which would influence the concentration variation with time of a second and/or third antibacterial component in a mixture susceptible to such drug efflux. Again this could provide valuable information to underpin further multi-action antibacterial design approaches.

Triple combination therapy was also advanced to clinical trials with the combination of imipenem, cilastatin and relebactam for the treatment of infections due to Gram-negative bacteria (Domalaon et al. 2018; Butler et al. 2017). This combination (Recarbrio) has now been approved by the U.S. Food and Drug Administration for the treatment of particular Gram-negative infections (Papp-Wallace 2019). In this combination, degradation of the carbapenem antibacterial, imipenem, is mitigated by the adjuvants cilastatin (a dehydropeptidase I inhibitor) and the β-lactamase inhibitor relebactam (Fig. 2.12).

Furthermore, cilastatin is a proximal tubule uptake blocker of protonated antibiotics, affording further amelioration of possible kidney damage with this combination. However, some β-lactamases attack imipenem, particularly metallo β-lactamases including imipenases (Hong et al. 2015), thus posing a restriction on the spectrum of use. From this knowledge one might then consider incorporating another wide spectrum β-lactamase inhibitor in place of relebactam to help advance this approach. Of particular value in this connection might be the inclusion in the combination of one of the bicyclic boronates with inhibitory activity against both serine-β-lactamases and metallo-β-lactamases (Krajnc et al. 2019). A useful recent

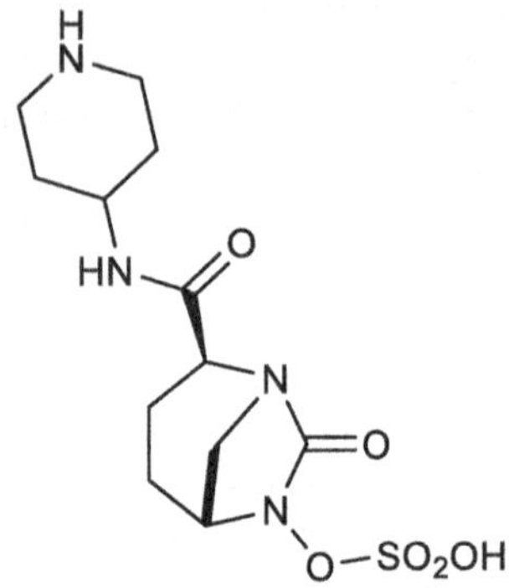

Fig. 2.12 Structure of the β-lactamase inhibitor relebactam

review also summarises a large amount of work over the past decade on mechanistic and structural aspects informing the design of inhibitors for the particularly problematic New Delhi metallo-lactamase (NDM-1) enzyme (Linciano et al. 2019).

The triple combination of meropenem-piperacillin-tazobactam is a synergistic one which suppresses resistance in MRSA (Gonzales et al. 2015). Bush (2015) has published a commentary on this paper including a clear mechanism of action representation in which tazobactam inhibits the staphylococcal penicillinase (Bla) preventing it from hydrolysing the PBP2-binder, piperacillin. The third component, meropenem, binds to PBP1 inhibiting transpetidation, as well at the allosteric site of PBP2a and opening the active site for β-lactam binding. As a result of these binding processes cell wall synthesis is inhibited.

A similar triple drug combination has been described involving a quinazolinone allosteric inhibitor of PBP 2a (Fig. 2.13, R an ethynyl group), which synergises the action of the second component piperacillin; the third component was again the β-lactamase inhibitor tazobactam. This combination showed good good activity and was efficacious in vivo in a mouse model utilizing methicillin-resistant *Staphylococcus aureus* (Janardhanan et al. 2019). Similar activity with a nitrile analogue (Fig. 2.13, R=CN) was also seen.

A three component drug combination to treat *Mycobacterium tuberculosis*, the NiX-TB study, using orally administered bedaquiline, pretomanid and high-dose

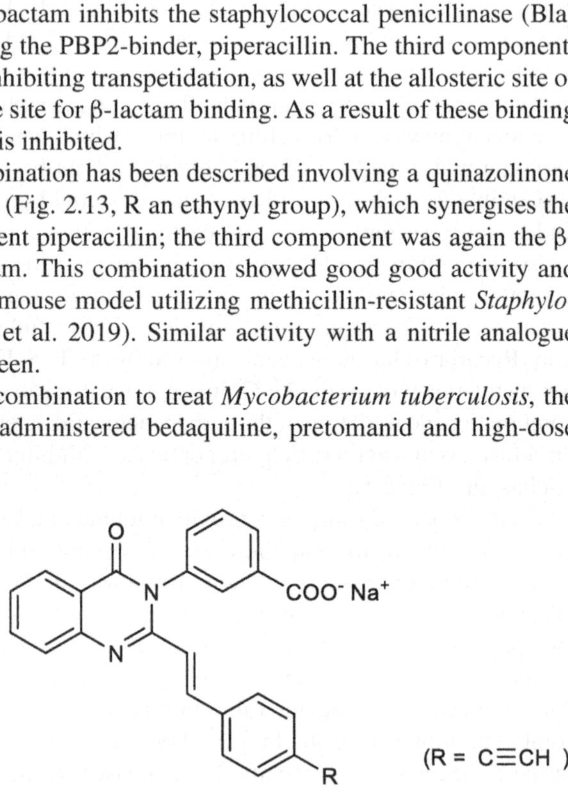

Fig. 2.13 Structure of a quinazolinone-based PBP allosteric site binder

linezolid, has shown great promise in its effectiveness (Vjecha et al. 2018; Conradie et al. 2020) and is proceeding to commercialisation (TB Alliance and Mylan 2019). Another three drug combination of clarithromycin, the proton pump inhibitor omeprazole, and amoxicillin has been used for the treatment of *Helicobacter pylori* but resistance to clarithromycin is now developing making this treatment less effective and extension to a four drug combination now looks a better treatment regime for *Helicobacter pylori* as noted in Sect. 2.1.4.1 (Malfertheiner et al. 2011).

It should be noted that every component in a three component mixture does not have to show antibacterial activity. Illustrative of this, Ferrer-Espada et al. (2019) have shown that the permeability-increasing antimicrobial peptide polymixin B nonapeptide can synergize with a MexAB-OprM efflux pump system inhibitor (particularly with phenylalanine-arginine-naphthylamide, PAβN) when the two are co-administered with the antibiotic azithromycin in planktonic and biofilm forming cells of *Pseudomonas aeruginosa* resulting in a significant boost in potency of the azithromycin in multi-drug resistant strains of this bacterium. The sensitivity to other antibiotics could also be changed in this way (Ferrer-Espada et al. 2019).

Administration

There are potential complications in the administration of triple combinations of drugs in terms of the physical requirement for three separate tablets or in combining three agents in two tablets one containing two of the drugs and the other just one, or all three drugs in a single specially constructed tablet or in liposomes or nanoparticles. There are still a number of considerations though which need to be addressed with tabletted combinations including the potential for mutual interference with absorption, metabolic clearance or modes of action, together with ensuring there is not too large a mismatch with dosage schedules of the combination components (Lowe 2018).

After administration, account needs to be taken of pharmacokinetic differences between the three components. One of the significant potential problems with oral antibacterial therapies based on drug combinations is, that in vivo, the different drugs are likely to have different absorption rates and different pharmacokinetic profiles. This in turn has the potential to compromise antibacterial potency since different concentrations of drugs could be present at the bacterial target sites at different times. So ways to maintain efficacious concentrations at each target site can be an issue and thus incorporating molecular features in each drug to address these issues needs to be considered. In addition, with triple combinations the likelihood of undesirable off-target effects is increased.

2.1.3.2 Antibacterial Photodynamic Therapy (aPDT)

$$\mathbf{A + B + light}$$

This special combination, which forms the basis of photodynamic therapy, is a variation on type (i) A + B + C triple or more actions but is somewhat different in that it requires light to complete the production of a third active species for attack on multiple potential bacterial sites. Light is not a third 'molecular entity C' as such but a substitute for it and a necessary third 'component' as an activating agent. One compound in the mixture is triplet oxygen (A) as a 'prodrug' which is converted to the active singlet oxygen through visible light irradiation in the presence of a sensitiser (B) and energy transfer (Type II mechanism), or to other cytotoxic reactive oxygen species such as superoxide (a radical anion) or free radicals (for example hydroxyl radicals) through charge transfer (Type I mechanism). These reactive species have potentially three or possibly more actions through reactions with susceptible sites in bacterial proteins, lipids or nucleic acids. Antibacterial photodynamic therapy (aPDT) is a significant research area and a good review of the field is presented in Liu et al. (2015). Dharmaratne et al. (2020) have reviewed a PDT for MRSA and general outstanding issues or problems are covered at the end of this review.

Light can also activate in other ways for example through reversible switching (Klaue et al. 2018), photoisomerisation, light induced electrocyclic reactions, through alterations in molecular shape and polarity, and also through phototriggered targeting with nanomedicine applications (Arrue and Ratjen 2017; Velema et al. 2014 see also Sect. 4.3.3 for photopharmacology).

A further useful variation on the light-activated combination theme is to add other functions to the photosensitising molecule to increase the likelihood of more potent antibacterial outcomes. For example the sensitiser may be converted into a dual function hybrid-photosensitiser and efflux pump evader/blocker (B − C) and this variation could then be expressed as A + (B − C) + light, where A is triplet oxygen. This combination has been developed in the work of Rineh et al. (2017, 2018). One of the B − C hybrids they described (Fig. 2.14) was based on the photosensitiser methylene blue (Fig. 2.18b) and incorporated a terminal 5-nitro-2-arlyindolic moiety to mimic the NorA efflux pump inhibitor INF-55. Enhancement (relative to methylene blue) of the inactivation of MRSA in vitro and in vivo (murine model) was seen with this hybrid (Rineh et al. 2017). Enhanced activity against two Gram-negatives *Escherichia coli* and *Acinetobacter baumannii* with this and related hybrids relative to methylene blue was also seen even though these bacteria do not express the NorA efflux pump (Rineh et al. 2017).

This approach of incorporating other orthogonal pharmacophoric features into the photosensitiser component affords further opportunities for triple or higher

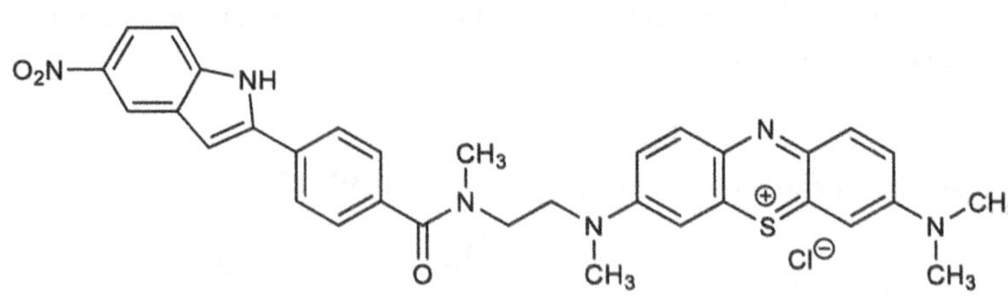

Fig. 2.14 Structural representation of the hybrid methylene blue-efflux pump inhibitor INF55-(Ac)en-MB 2

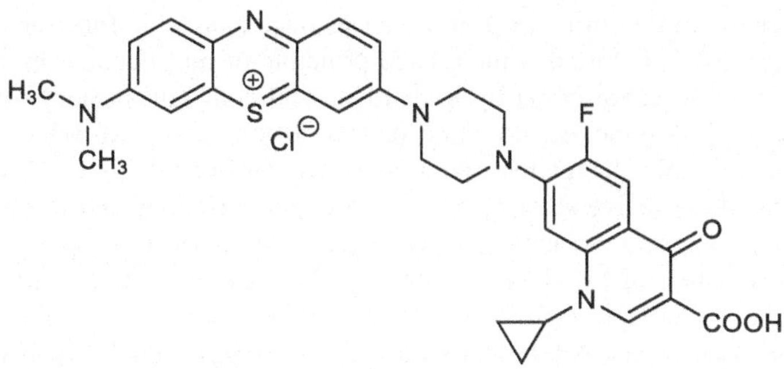

Fig. 2.15 Structure of a suggested methylene blue-ciprofloxacin hybrid

action design protocols. For example, the photosensitiser might incorporate units to bind to key bacterial enzymes, or have bacterial DNA binding capability together with features to enhance both Gram-negative and Gram-positive penetration. One suggested expression of this may be a methylene blue-ciprofloxacin hybrid with one ciprofloxacin group attached directly via the piperidine substituent, leaving one NMe_2 group on the methylene blue (Fig. 2.15). This hybrid might localise in the DNA area through the ciprofloxacin moiety and then on light exposure generate singlet oxygen for further DNA damage amongst other outcomes. Attack on the ciprofloxacin moiety may be a problem with singlet oxygen so perhaps a time-delay before light exposure would be warranted here. Also energy transfer to the ciprofloxacin group (quenching) may occur rather than to triplet oxygen, depending on the relative energy gaps. The bulky ciprofloxacin group could have a positive effect though in helping the methylene blue portion to evade the efflux pumps.

Another interesting aspect here is that of bacterial efflux pumps also acting as influx pumps or transporters as well as efflux vehicles. The recent paper by Jindal et al. (2019) points to a number of transporters in *Escherichia coli* which are involved in the influx/efflux of two cationic dyes (the carbocyanine diS-C3(5) and SYBR Green). Diffusional phospholipid bilayer transport is negligible contrary to the general belief that diffusion is the major means of ingress with protein-based transporters being of lesser significance in most cases. This does raise the possibility of designing influx substrates which, after being taken in to the bacterial cell, are then converted, possibly by a light-induced process, into another active species but one which evades or blocks efflux. The light used might then have two functions-singlet oxygen generation as well as perhaps light induced E/Z isomerisation of another substituent group containing an azo or ethenyl moiety on the methylene blue which changes the molecular shape sufficiently to elude efflux. This design principle of doing two things with one light source has considerable attraction. Other changes intracellularly might be non-light mediated, however, and could result from exposure to different enzymes. For example intracellular peptide deformylase-mediated hydrolysis of a formamide unit to a free amino group susceptible to protonation could be advantageous, prior to light exposure. Intramolecular non-covalent interactions might also come into play to change shape or conformational preferences once a suitable functional group is exposed. It is worth noting that the related principle of multi-action reinforcement has been used with great effect by Richardson and colleagues (Stacy et al. 2016) in the development of the dipyridyl thiosemicarbazone-based anti-cancer agents. In the lysosome, the pH change to around 5 was crucial in avoiding efflux as well as allowing for the effective scavenging of free copper (II) ions and the subsequent redox cycling to generate reactive oxygen species. Antibacterial design possibilities involving utilization of pH changes within the bacterial cell might also be further explored.

Berberine is a good DNA binder so perhaps it may potentially be beneficial to look at designing modified photosensitisers with similar overall shape and charge characteristics to berberine while avoiding the presence of reactive benzylic hydrogens. In an expression of this idea with the Occam's Razor principle in mind—"don't include more than is necessary"—it might be worthwhile considering achieving the bent shape with a fused efflux pump blocker component while still preserving the required photosensitisation activity. One could also consider not fusing the pump blocker but joining it to the phenothiazine by a single C–C single bond connection for example from the indole 2- or 3-position to the 7-position of the phenothiazine (Fig. 2.16b).

A further intriguing design possibility might be to convert the berberine skeleton itself into a potential triplet photosensitiser as indicated in the proposed structure shown in Fig. 2.17a. It is still likely to bind to bacterial DNA and then on exposure to visible light generate singlet oxygen and then reactive oxygen species near the DNA (multiple action). If it is still an efflux pump substrate a potential pump inhibitor moiety (or evader) could be added in group A. Berberine is a good starting point or template for such design in view of the range of atoms and rings inherent in its structure. This then allows for many ways to vary the positions and nature of the skeletal atoms as well as the substituents. In considering these variations it is recommended that one not be constrained initially by synthetic feasibility so as not to exclude potentially viable candidates thus maximising new structural possibilities.

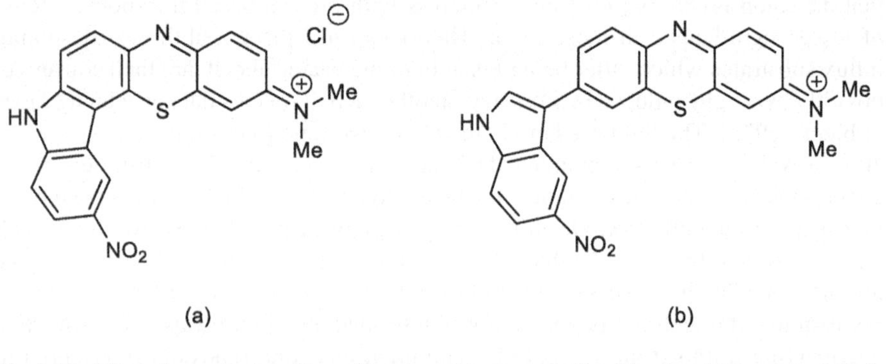

Fig. 2.16 Alternative methylene blue-efflux pump blocker hybrids with a fused (**a**) or joined (**b**) 5-nitroindolic component

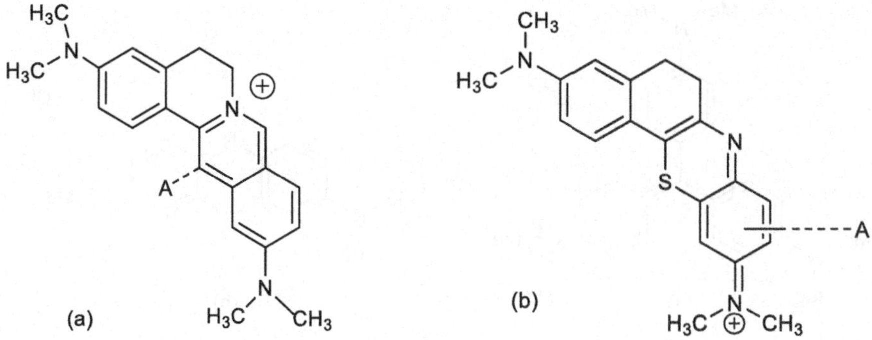

Fig. 2.17 Potential triplet photosensitiser based on the berberine core (**a**) and a thia analogue (**b**)

Synthetic considerations can be incorporated in later refinements. For example one might initially consider moving the nitrogen by one atom position and substituting carbon by sulfur in the same ring as in the theoretical structure (Fig. 2.17b); the added pharmacophore A might then be attached to another ring.

One does need to consider a range of parameters in any photosensitiser for use in the likely efficient production of singlet oxygen (see DeRosa and Crutchley 2002 for a good review on this). An important parameter is photostability and this will impinge on the nature of A and the way it may be linked to the heterocyclic core (Fig. 2.17). Another key parameter involves energy transfer and, in this case, whether DNA-bound photosensitiser will still act as an efficient singlet sensitiser when exposed to visible light or not. If stability and sensitiser issues are not evident then these types of compounds should significantly potentiate bacterial inactivation.

Dyes used in the staining of bacteria might also provide significant possibilities for targeted production of singlet oxygen or reactive oxygen species in or close to the bacterial targets. For example basic fuchsin, crystal violet (Fig. 2.18a) (also known as gentian violet and used as a topical antimicrobial), methylene blue (Fig. 2.18b), and new analogues of these basic skeletons, could be considered. These compounds can target bacteria and on light exposure may generate a high concentration of reactive oxygen species on and exterior to the bacterial cell wall although unwanted damage to surrounding host tissues may be an issue. Mixed dye analogues like the suggested crystal violet-methylene blue hybrid analogue (Fig. 2.18c) might be worth investigating where the ring sulfur could also participate in charge delocalisation.

The synthesis of analogues of this suggested hybrid type have been discussed in Kanagasundaram et al. (2019), but these do not seem to include a *p*-dimethylamino substituent in the pendant aryl group. However, this would not seem to preclude synthesis of the *p*-dimethylaminophenylboronic acid precursor that would be required.

The effectiveness of PDT can be enhanced with added inorganic salts (Hamblin 2017) for example with potassium iodide, in which the iodide ion provides access to other antibacterial entities like the periodide ion or hydrogen peroxide and reactive

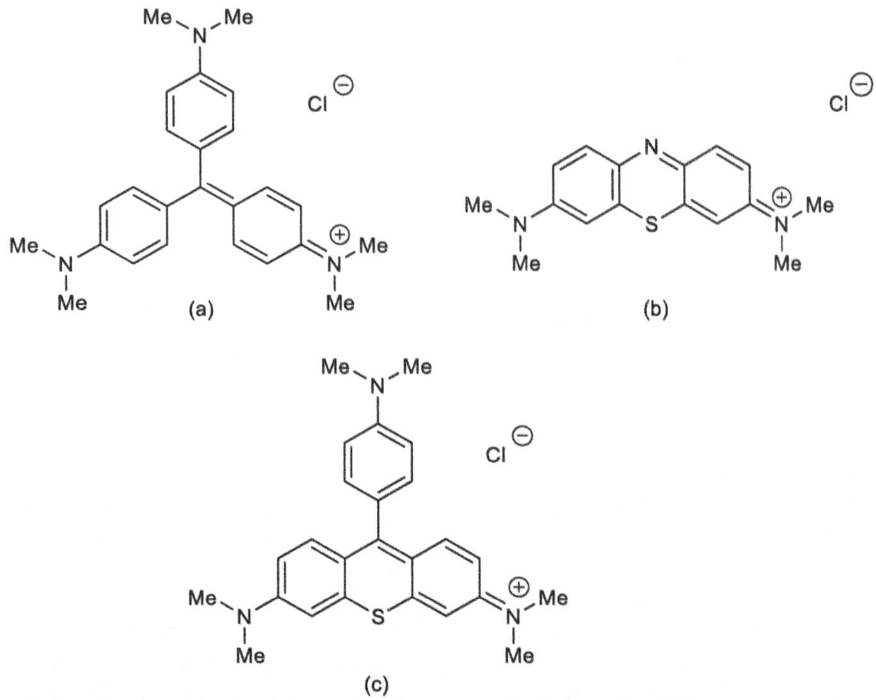

Fig. 2.18 Structures of crystal violet (**a**), methylene blue (**b**) and a hypothetical crystal violet-methylene blue hybrid analogue (**c**)

species like the peroxy radical or the iodine radical anion (Hamblin 2017; Huang et al. 2017). This represents an addition to the combination design approach which could be applied in principle to multiply active antibacterial design.

Also it has been shown that incorporating noble metal nanoparticles with the dye photosensitizer significantly improves bacteriocidal potency, possibly as a result of improved photosensitizer localisation and a plasmonic enhancement effect of the metal nanoparticles (Zhang 2018).

Sonosenitizers, including methylene blue and fluoroquinolones like ciprofloxacin, may also generate reactive oxygen species (Chen et al. 2014). This implies some interesting new combination possibilities such as the methylene blue-ciprofloxacin derivative (Fig. 2.15), ground state oxygen, and focussed ultrasound which can penetrate human tissue over 10 cm, which is much greater than for visible light or near-IR penetration. The reactive oxygen species generated may then synergistically interact with DNA bases near the ciprofloxacin binding sites in gyrase B and topoisomerase IV.

Combination treatment of skin and mucosal infections with aPDT and antibacterial compounds has been reviewed, and even though only a limited number of clinical case studies were assessed, it appeared that additive or synergistic effects could potentially ensue to manage these types of infections. The most effective combinations involved 5-aminolevulinic acid or phenothiazinium dyes in the aPDT with the antibacterials administered post-aPDT consistent with PDT damage of the bacterial cell wall or membrane then enabling increased penetration of the antibacterial agent (Pérez-Laguna et al. 2019).

Instead of a combination approach, in other related work, a rose-bengal-antimicrobial peptide conjugate (single molecule hybrid) which is activated by ultrasound (ROSs produced) was shown to be effective against Gram-negative and Gram-positive bacteria (Costley et al. 2017). Extension of this to aPDT should also be possible.

2.1.3.3 Type (II) Triple Combination (A − B) + C + D

In this combination, a dual action hybrid (A − B) or related prodrug plus two separate compounds (C and D) could be used. A number of variations on this theme could be envisaged including two dual acting hybrids and a third single component with one further activity. No specific examples have been found in the literature but the type is included to stimulate further consideration as a possible triple combination design option for the future.

2.1.4 Quadruple Combinations with Four or More Actions

2.1.4.1 Type (I) Combination A + B + C + D

A recent paper by Yeh and co-workers (Tekin et al. 2018) reports on the discovery of potent four (and five) antibacterial combinations against *Escherichia coli*. They used eight antibiotics in the studies which had varying modes of action. These antibiotics were ampicillin, cefoxitin sodium salt, trimethoprim, ciprofloxacin hydrochloride, streptomycin, doxycycline hydrate, erythromycin and the sodium salt of fusidic acid. Every possible four- and five- compound combination (a total of 18,278 combinations) including a significant number of combinations with varying dosages, were assessed experimentally in vitro. All the two- and three-drug combinations were tested as well. Surprisingly, a large number of four drug combinations (1676 out of 4007) were more effective than predicted on the independent individual antibiotic effects in stopping the growth of *Escherichia coli*, while 6443 five-drug combinations (out of 11,642) were more effective. An increased occurrence of emergent antagonism was also seen with the four or five way combinations. As noted by Coates (2019) this type of work on antibacterial combinations has important implications for future therapy. It is also important with regard to informing structural designs for multi-active hybrids (or preferably prodrugs), but trying to incorporate four or five separate activity capabilities in a single molecule has major challenges.

A combination of four drugs, isoniazid, rifampicin, ethambutol and pyrazinamide, has been used as a standard of care for infections caused by *Mycobacterium tuberculosis* (Worthington and Melander 2013). Another treatment regime (BPaMZ) for TB involving four drugs—bedaquiline, pretomanid, moxifloxacin and pyrazinamide—began in November 2014 (Vjecha et al. 2018). This evaluation study has now progressed to an on-going late-stage SimpliciTB clinical trial for the treatment of drug-sensitive, and multi drug resistant, TB; very promising results were obtained with the BPaMZ regimen in an earlier Phase IIb study (TB Alliance and Mylan 2019). But while four or more drugs are used one still needs to show there is a genuine combination effect of the four (or more).

The pathogen *Helicobacter pylori* has also been the focus of multi-drug therapy and one of these was a four drug combination administered as a single capsule with three compounds plus omeprazole. This treatment regime for *Helicobacter pylori* was found to be more effective in eradicating the bacterium compared with the gold standard triple combination (Malfertheiner et al. 2011). Coming back to one of the themes in this book, a meta-analysis of randomized clinical trial data has shown that the antibacterial alkaloid berberine is beneficial in quadruple therapy for *Helicobacter pylori* in China, when used in combination with standard triple therapy incorporating the antibiotics clarithromycin and amoxicillin together with a proton pump inhibitor as an anti-secretory agent. This then appeared to be a promising protocol to improve the eradication rate for this bacterium while reducing adverse events and promoting ulcer healing (Hu et al. 2020).

2.1.4.2 Type (II) Combination (A − B) + C + D + E

This is a variation on the four-component type (i) theme which includes the incorporation of dual or triple action hybrids (A − B) (or related prodrugs) plus three at least single action drugs thus potentially resulting in more than four interactions or actions. However the author is not aware of any specific realisations of this antibacterial combination type as yet but it would be of interest to pursue. One might note that four component systems form the basis of biology and our existence, namely four different DNA bases rather than say two. Four gives ample opportunity for variety and functional expressions. A similar power could be unleashed looking at four possible synchronous or near synchronous interactions involving a number of bacterial targets.This space could be fertile territory for further innovative developments.

2.1.5 Pentuple Combinations

Five component drug combinations (A + B + C + D + E) were also included in the work reported by Tekin et al. (2018) in the previous Sect. 2.1.4.1. At a clinical level, five separate drugs are being used in the TB trial, endTB, which began in December

2016 (Vjecha et al. 2018). In theory, extensions to include one or more dual or triple action hybrids (and/or associated prodrugs) in the five component mixture would extend the activity profile beyond five separate single actions, but it is considered less likely to be realised in practice though.

2.2 Summary

The investigation of the antibacterial activities of double, triple and higher component combinations, and variations including aPDT, continues to flourish and the likelihood of further applications in the clinic is high. As well as this, the increased knowledge of combination treatments with two or more compounds, especially if synergistic, will be useful in better informing the intentional design of potentially multi-action single molecule agents. Subsequent chapters will concentrate on such single molecule design principles, particularly those that incorporate new or untried proposals.

References

Abdali N, Parks JM, Haynes KM et al. (2017) Reviving antibiotics: efflux pump inhibitors that interact with AcrA, a membrane fusion protein of the AcrAB-TolC multidrug efflux pump. ACS Infect Dis 3 (1):89–98

Aboulmagd E, Alsultan AA (2014) Synergic bactericidal activity of novel antibiotic combinations against extreme drug resistant *Pseudomonas aeruginosa* and *Acinetobacter baumannii*. Afric J Microbiol Res 8 (9): 856–861

Abrudan MI, Smakman F, Grimbergen AJ et al. (2015) Socially mediated induction and suppression of antibiosis during bacterial coexistence. Proc Natl Acad Sci U S A 112 (35):11054–11059

Ammeter D, Idowu T, Zhanel GG et al. (2019) Development of a nebramine-cyclam conjugate as an antibacterial adjuvant to potentiate β-lactam antibiotics against multidrug-resistant *P. aeruginosa*. J Antibiot (Tokyo) 72 (11):816–826

Arrue L, Ratjen L (2017) Internal targeting and external control: Phototriggered targeting in nanomedicine. ChemMedChem 12:1908–1916

Ball AR, Casadei G, Samosorn S et al. (2006) Conjugating berberine to a multidrug resistance pump inhibitor creates an effective antimicrobial. ACS Chem Biol 1 (9):594–600

Beppler C, Tekin E, Mao Z et al. (2016) Uncovering emergent interactions in three-way combinations of stressors. J R Soc Interface 13:20160800

Bezar IF, Mashruwala AA, Boyd JM et al. (2019) Drug-like fragments inhibit agr-mediated virulence expression in *Staphylococcus aureus*. Sci Rep 9 (1):6786

Bremner JB, Ambrus JI, Samosorn S (2007) Dual-action based approaches to antibacterial agents. Curr Med Chem 14(13):1459–1477

Brochado AR, Telzerow A, Bobonis J et al. (2018) Species-specific activity of antibacterial drug combinations. Nature 559 (7713):259–263

Brown DG, Boström J (2018) Where do recent small molecule clinical development candidates come from? J Med Chem 61 (21):9442–9468

Brown ED, Wright GD (2016) Antibacterial drug discovery in the resistance era. Nature 529:336–343

Bush K (2015) Antibiotics: synergistic MRSA combinations. Nat Chem Biol 11 (11):832–833

Butler MS, Paterson DL (2020) Antibiotics in the clinical pipeline in October 2019. J Antibiot (Tokyo) 73 (6):329–364

Butler MS, Blaskovich MA, Cooper MA (2017) Antibiotics in the clinical pipeline at the end of 2015. J Antibiot (Tokyo) 70 (1):3–24

Conradie F, Diacon AH, Ngubane N et al. (2020) Treatment of highly drug-resistant pulmonary tuberculosis. N Eng J Med 382(10):893–902

Calvert MB, Jumde VR, Titz A (2018) Pathoblockers or antivirulence drugs as a new option for the treatment of bacterial infections. Beilstein J Org Chem 14:2607–2617

Chen H, Zhou X, Gao Y et al. (2014) Recent progress in development of new sonosensitizers for sonodynamic cancer therapy. Drug Discov Today 19 (4):502–509

Chou TC (2010) Drug combination studies and their synergy quantification using the Chou-Talalay method. Cancer Res 70 (2):440–446

Clatworthy AE, Pierson E, Hung DT (2007) Targeting virulence: a new paradigm for antimicrobial therapy. Nat Chem Biol 3(9):541–548

Coates A (2019) The future of antibiotics lies in combination treatments. Future Drug Discov 1(1):FDD5

Costley D, Nesbitt H, Ternan N et al. (2017) Sonodynamic inactivation of Gram-positive and Gram-negative bacteria using a Rose Bengal-antimicrobial peptide conjugate. Int J Antimicrob Agents 49 (1):31–36

Crow JM (2019) The fifth element. Chem World 16(5): 48–51

Decuyper L, Jugic M, Sosic I et al. (2018) Antibacterial and β-lactamase inhibitory activity of monocyclic β-lactams. Med Res Rev 38 (2):426-503

DeRosa MC, Crutchley RJ (2002) Photosensitized singlet oxygen and its applications. Coord Chem Rev 233-234:351-371

Dharmaratne P, Sapugahawatte DN, Wang B et al. (2020) Contemporary approaches and future perspectives of antibacterial photodynamic therapy (aPDT) against methicillin-resistant *Staphylococcus aureus* (MRSA): A systematic review. Eur J Med Chem 200:112341

Domalaon R, Idowu T, Zhanel GG et al. (2018) Antibiotic hybrids: the next generation of agents and adjuvants against Gram-negative pathogens? Clin Microbiol Rev 31 (2):e00077–17

Domalaon R, Ammeter D, Brizuela M et al. (2019) Repurposed antimicrobial combination therapy: Tobramycin-ciprofloxacin hybrid augments activity of the anticancer drug mitomycin C against multidrug-resistant Gram-negative bacteria. Front Microbiol 10: Article 1556

Dickey SW, Cheung GYC, Otto M (2017) Different drugs for bad bugs: antivirulence strategies in the age of antibiotic resistance. Nat Rev Drug Discov 16 (7):457–471

El-Halfawy OM, Czarny TL, Flannagan RS et al. (2020) Discovery of an antivirulence compound that reverses β-lactam resistance in MRSA. Nat Chem Biol 16 (2):143–149

Ferrer-Espada R, Shahrour H, Pitts B et al. (2019) A permeability-increasing drug synergizes with bacterial efflux pump inhibitors and restores susceptibility to antibiotics in multi-drug resistant *Pseudomonas aeruginosa* strains. Sci Rep 9 (1):3452

Fontaine F, Hequet A, Voisin-Chiret A-S et al. (2014) First identification of boronic species as novel potential inhibitors of the *Staphylococcus aureus* NorA efflux pump. J Med Chem 57 (6):2536–2548

Gomes-Fernandes M, Laabei M, Pagan N et al. (2017) Accessory gene regulator (Agr) functionality in *Staphylococcus aureus* derived from lower respiratory tract infections. PloS One 12 (4):e0175552

Gonzales PR, Pesesky MW, Bouley R et al. (2015) Synergistic, collaterally sensitive β-lactam combinations suppress resistance in MRSA. Nat Chem Biol 11 (11):855–861

Greenberg M, Kuo D, Jankowsky E et al. (2018) Small-molecule AgrA inhibitors F12 and F19 act as antivirulence agents against Gram-positive pathogens. Sci Rep 8 (1):14578

Gutierrez A, Jain S, Bhargava P et al. (2017) Understanding and sensitizing density-dependent persistence to quinolone antibiotics. Mol Cell 68(6):1147–1154.e3

Hamblin MR (2017) Potentiation of antimicrobial photodynamic inactivation by inorganic salts. Expert Rev Anti Infect Ther 15 (11):1059–1069

Hamoud R, Zimmermann S, Reichling J et al. (2014) Synergistic interactions in two-drug and three-drug combinations (thymol, EDTA and vancomycin) against multi drug resistant bacteria including *E. coli*. Phytomedicine 21 (4):443–447

Haynes KM, Abdali N, Jhawar V et al. (2017) Identification and structure-activity relationships of novel compounds that potentiate the activities of antibiotics in *Escherichia coli*. J Med Chem 60 (14):6205–6219

Hecker SJ, Reddy KR, Lomovskaya O et al. (2020) Discovery of cyclic boronic acid QPX7728, an ultrabroad-spectrum inhibitor of serine and metallo-β-lactamases. J Med Chem 63 (14):7491–7507

Hecker SJ, Reddy KR, Totrov M et al. (2015) Discovery of a cyclic boronic acid β-lactamase inhibitor (RPX7009) with utility vs class A serine carbapenemases. J Med Chem 58 (9):3682–3692

Hernandez V, Akama T, Li X et al. (2016) Boron in medicinal chemistry. In: Desai MC (ed -in-chief) 2016 Medicinal chemistry reviews, vol 51. Medicinal Chemistry Division of the American Chemical Society, Washington, pp 329–345

Hong DJ, Bae IK, Jang IH et al. (2015) Epidemiology and characteristics of metallo-β-lactamase-producing *Pseudomonas aeruginosa*. Infect Chemother 47 (2):81–97

Hu Q, Peng Z, Li L et al. (2020) The efficacy of berberine-containing quadruple therapy on *Helicobacter pylori* eradication in China: A systematic review and meta-analysis of randomized clinical trials. Front Pharmacol 10 (1694)

Huang L, Szewczyk G, Sarna T et al. (2017) Potassium iodide potentiates broad-spectrum antimicrobial photodynamic inactivation using Photofrin. ACS Infect Dis 3:320–328

Janardhanan J, Bouley R, Martínez-Caballero S et al. (2019) The quinazolinone allosteric inhibitor of PBP 2a synergizes with Piperacillin and Tazobactam against Methicillin-resistant *Staphylococcus aureus*. Antimicrob Agents Chemother 63 (5):e02637–18

Jia J, Zhu F, Ma X et al. (2009) Mechanisms of drug combinations: interaction and network perspectives. Nat Rev Drug Discov 8 (2):111–128

Jindal HM, Zandi K, Ong KC et al. (2017) Mechanisms of action and in vivo antibacterial efficacy assessment of five novel hybrid peptides derived from indolicidin and ranalexin against *Streptococcus pneumoniae*. PeerJ 5:e3887, doi: https://doi.org/10.7717/peerj.3887

Jindal S, Yang L, Day PJ et al. (2019) Involvement of multiple influx and efflux transporters in the accumulation of cationic fluorescent dyes by *Escherichia coli*. BMC Microbiol 19 (1):195

Kanagasundaram T, Timmermann A, Kramer CS et al. (2019) A new approach to silicon rhodamines by Suzuki-Miyaura coupling - scope and limitations. Beilstein J Org Chem 15:2569–2576

Klaue K, Garmshausen Y, Hecht S (2018) Taking photochromism beyond visible: direct one-photon NIR photoswitches operating in the biological window. Angew Chem Int Ed Engl 57 (5):1414–1417

Kotani K, Matsumura M, Morita Y et al. (2019) 13-(2-Methylbenzyl)berberine is a more potent inhibitor of MexXY-dependent aminoglycoside resistance than Berberine. Antibiotics (Basel) 8 (4):212

Krajnc A, Lang PA, Panduwawala TD et al. (2019) Will morphing boron-based inhibitors beat the β-lactamases? Curr Opin Chem Biol 50:101–110

Laudadio E, Cedraro N, Mangiaterra G et al. (2019) Natural alkaloid berberine activity against *Pseudomonas aeruginosa* MexXY-mediated aminoglycoside resistance: in silico and in vitro studies. J Nat Prod 82 (7):1935–1944

Lee Y, Kim J, Trinh S (2019) Meropenem-Vaborbactam (Vabomere^{TM}): Another option for carbapenem-resistant Enterobacteriaceae. Pharm Ther (P&T$_{®}$) 44(3):110–113

Lewis K (2020) The science of antibiotic discovery. Cell 181 (1):29–45

Linciano P, Cendron L, Gianquinto E et al. (2019) Ten years with New delhi Metallo-β-lactamase-1 (NDM-1): from structural insights to inhibitor design. ACS Infect Dis 5 (1):9–34

Liu Y, Qin R, Zaat SAJ et al. (2015) Antibacterial photodynamic therapy: Overview of a promising approach to fight antibiotic-resistant bacterial infections. J Clin Transl Res 1(3):140–167

Liu B, Trout REL, Chu G-H et al. (2020) Discovery of taniborbactam (VNRX-5133): A broad-spectrum serine- and metallo-β-lactamase inhibitor for carbapenem-resistant bacterial infections. J Med Chem 63 (6):2789–2801

Liu B, Hamrick J, Daigle DM et al. (2019) Exploiting penicillin-binding protein-mediated inhibition of cell biosynthesis for the treatment of bacterial infections. In: Bronson JJ (ed-in-chief) 2019 Medicinal chemistry reviews, vol 54. Medicinal Chemistry Division of the American Chemical Society, Washington, pp 253–269

Lowe D (2018) Cracking the combination. Chem World 15 (5):22

Malfertheiner P, Bazzoli F, Delchier J-C et al. (2011) *Helicobacter pylori* eradication with a capsule containing bismuth subcitrate potassium, metronidazole, and tetracycline given with omeprazole versus clarithromycin-based triple therapy: A randomised, open-label, non-inferiority, phase 3 trial. Lancet 377:905-913

Nguyen LP, Pinto NA, Vu TN et al. (2020) In vitro activity of a novel Siderophore-Cephalosporin, GT-1 and Serine-Type β-Lactamase inhibitor, GT-055, against *Escherichia coli*, *Klebsiella pneumoniae* and *Acinetobacter* spp. panel strains. Antibiotics (Basel) 9 (5):267

Ocampo PS, Lázár V, Papp B et al. (2014) Antagonism between bacteriostatic and bactericidal antibiotics is prevalent. Antimicrob Agents Chemother 58 (8):4573–4582

O'Dwyer K, Spivak AT, Ingraham K et al. (2015) Bacterial resistance to leucyl-tRNA synthetase inhibitor GSK2251052 develops during treatment of complicated urinary tract infections. Antimicrob Agents Chemother 59 (1):289–298

Oldfield E, Feng X (2014) Resistance-resistant antibiotics. Trends Pharmacol Sci 35 (12):664–674

Papp-Wallace KM (2019) The latest advances in beta-lactam/beta-lactamase inhibitor combinations for the treatment of Gram-negative bacterial infections. Expert Opin Pharmacother 20 (17):2169–2184

Pedrola M, Jorba M, Jardas E et al. (2019) Multicomponent reactions upon the known drug Trimethoprim as a source of novel antimicrobial agents. Front Chem 7:475

Pérez-Laguna V, Gilaberte Y, Millán-Lou MI et al. (2019) A combination of photodynamic therapy and antimicrobial compounds to treat skin and mucosal infections: a systematic review. Photochem Photobiol Sci 18 (5):1020–1029

Pham TDM, Ziora ZM, and Blaskovich MAT (2019) Quinolone antibiotics. MedChemComm 10:1719-1739

Prati F, Uliassi E, Bolognesi ML (2014) Two diseases, one approach: multitarget drug discovery in Alzheimer's and neglected tropical diseases. MedChemComm 5 (7):853–861

Ramaswamy VK, Cacciotto P, Malloci G et al. (2017) Computational modelling of efflux pumps and their inhibitors. Essays Biochem 61 (1):141–156

Rineh A, Dolla NK, Ball AR et al. (2017) Attaching the NorA efflux pump inhibitor INF55 to methylene blue enhances antimicrobial photodynamic inactivation of methicillin-resistant *Staphylococcus aureus* in vitro and in vivo. ACS Infect Dis 3 (10):756–766

Shaw E, Wuest WM (2020) Virulence attenuating combination therapy: a potential multi-target synergy approach to treat *Pseudomonas aeruginosa* infections in cystic fibrosis patients. RSC Medicinal Chemistry 11 (3):358–369

Simões M, Bennett RN, Rosa EA (2009) Understanding antimicrobial activities of phytochemicals against multidrug resistant bacteria and biofilms. Nat Prod Rep 26 (6):746–757

Stacy AE, Palanimuthu D, Bernhardt PV et al. (2016) Structure-activity relationships of Di-2-pyridylketone, 2-Benzoylpyridine, and 2-Acetylpyridine thiosemicarbazones for overcoming Pgp-mediated drug resistance. J Med Chem 59 (18):8601–8620

Stokes JM, MacNair CR, Ilyas B et al. (2017) Pentamidine sensitizes Gram-negative pathogens to antibiotics and overcomes acquired colistin resistance. Nat Microbiol 2: 17028

Su F, Wang J (2018) Berberine inhibits the MexXY-OprM efflux pump to reverse imipenem resistance in a clinical carbapenem-resistant *Pseudomonas aeruginosa* isolate in a planktonic state. Exp Ther Med 15 (1):467–472

Sully EK, Malachowa N, Elmore BO et al. (2014) Selective chemical inhibition of agr quorum sensing in *Staphylococcus aureus* promotes host defense with minimal impact on resistance. PLoS Pathog 10 (6):e1004174

TB Alliance and Mylan (2019) TB Alliance and Mylan announce collaboration to commercialize pretomanid to treat tuberculosis (TB). News release 18 April 2019. https://www.eurekalert.org/pub_releases/2019-04/b-taa041719.php. Accessed 16 December 2020

Tekin E, Beppler C, White C et al. (2016) Enhanced identification of synergistic and antagonistic emergent interactions among three or more drugs. J R Soc Interface 13:20160332

Tekin E, White C, Kang TM et al. (2018) Prevalence and patterns of higher-order drug interactions in *Escherichia coli*. NPJ Syst Biol Appl 4:31

Trombetta D, Castelli F, Sarpietro MG et al. (2005) Mechanisms of antibacterial action of three monoterpenes. Antimicrob Agents Chemother 49 (6): 2474–2478

Tyers M, Wright GD (2019) Drug combinations: a strategy to extend the life of antibiotics in the 21st century. Nat Rev Microbiol 17 (3):141–155

Velema WA, Szymanski W, Feringa BL (2014) Photopharmacology: Beyond proof of principle. J Am Chem Soc 136:2178–2191

Venatorx Pharmaceuticals (2020) Cefepime-Taniborbactam. https://www.venatorx.com/cefepime-taniborbactam/. Accessed 27 April 2020

Vjecha MJ, Tiberi S, Zumla A (2018) Accelerating the development of therapeutic strategies for drug-resistant tuberculosis. Nat Rev Drug Discov 17 (9):607–608

Wang S, Setlow B, Setlow P et al. (2016) Uptake and levels of the antibiotic berberine in individual dormant and germinating *Clostridium difficile* and *Bacillus cereus* spores as measured by laser tweezers Raman spectroscopy. J Antimicrob Chemother 71 (6):1540–1546

Wood K, Nishida S, Sontag ED et al. (2012) Mechanism-independent method for predicting response to multidrug combinations in bacteria. Proc Natl Acad Sci U S A 109 (30):12254–12259

Worthington RJ, Melander C (2013) Combination approaches to combat multidrug-resistant bacteria. Trends Biotechnol 31 (3):177–184

Yang X, Ammeter D, Idowu T et al. (2019) Amphiphilic nebramine-based hybrids rescue legacy antibiotics from intrinsic resistance in multidrug-resistant Gram-negative bacilli. Eur J Med Chem 175:187–200

York A (2020) New drugs for the antibacterial pipeline? Nat Rev Microbiol 18 (2):61–61

Zhang P (2018) Weaponizing oxygen to kill infections and disease. Paper presented at the ACS national meeting and exposition, Boston, MA, USA, pp 19–23

Zuo GY, Li Y, Han J et al. (2012) Antibacterial and synergy of berberines with antibacterial agents against clinical multi-drug resistant isolates of methicillin-resistant *Staphylococcus aureus* (MRSA). Molecules 17 (9):10322–10330

Chapter 3
Single Molecule Non-cleavable Multiply Active Antibacterials

"Sometimes, less is more."
—William Shakespeare (attrib.).

Good things come in threes (Latin-"omne trium perfectum")

Abstract Multiply active non-cleavable antibacterials constitute a complex molecular design area. The emphasis in this chapter is on design parameters for small molecule triple or higher action single agents based on background information from drug combinations and dual action hybrids. Established multiple action agents or multi-targeting of bacterial sites by single ligands are discussed and then new structural possibilities are suggested illustrating the key design principles involved. A feature of this chapter is the use of a known antibacterial, the natural product berberine, as the starting point to illustrate potentially generalizable structural modifications for deliberate multiple action design. Cross referencing to sections of other chapters is included where appropriate. In this chapter there is a concentration on the medicinal chemistry aspects in new antibacterial design.

J. Bremner, *Multiple Action-Based Design Approaches to Antibacterials*,
https://doi.org/10.1007/978-981-16-0999-2_3

3.1 Introduction to General Design Considerations

An alternative to using combinations of separate compounds is that of a single drug compound whose structural features intentionally incorporate aspects of the key recognition elements from some or all of the separate compounds in a combination. Such hybrid 'doing-more-for-less' compounds can be an attractive option in multiple-mechanism-based antibacterial design (Silver 2007; East and Silver 2013). In these compounds the different molecular recognition features are designed to enable interactions with a number of biological target sites. This can be considered as directed multi-targeting with a single non-cleavable agent rather than a mixture of separate compounds: one compound, non-cleavable with two or more actions (Decker 2017; Bolognesi 2019).

There are, however, both advantages and disadvantages with hybrids over combinations. Both approaches are multi-targeting and thus offer the potential advantage of slowing bacterial resistance development. With a combination, the dosage of each drug administered can be tailored to the target for optimal inhibition or antagonism, but with hybrids there is a potential difficulty with specific bacterial target sites being in different cellular locations, for example if one site is extracellular and the other intracellular. However with more efficient delivery expected with a single compound compared with combinations of compounds, it should still be possible to achieve optimal or greater concentrations at each target to elicit the responses required. There are, however, significant challenges with regard to conflicting physicochemical property requirements with hybrids in general as noted in the analysis by Morphy and Rankovic (2006). For multi-action hybrid design it is even more important the physical distribution of cell components/targets is taken into account and this aspect is a significant factor in the design process.

A hybrid has different definitions depending on the context. In biology, a hybrid is a plant or animal that has been produced from two different types of plant or animal consistent with the Latin meaning of the word (*hybrida*—a crossbred animal). It can also be something that is a mixture of two very different things or broadened to a mixture of two or more things (Cambridge English Dictionary). Thus a mixture of three or more components in one molecule could still be referred to as a hybrid (Domalaon et al. 2018). Chemically, resonance hybrid structures may also reflect input from more than two hypothetical resonance contributing structures. In this book, though, the word is restricted to inclusion of a mixture of two or more target recognition features in a single molecule.

3.1.1 General Approaches to Hybrids

General approaches to hybrid design, covering key design points and principles, have been elaborated in various reviews and book chapters (Dolles and Decker 2017; Morphy 2012; Simoni et al. 2012; Brötz-Oesterhelt and Brunner 2008). Further work in the area with respect to bifunctional antimicrobials has been reviewed more recently by Klahn and Brönstrup (2017). These authors differentiate such bifunctional compounds into two general classes for dual targeting agents: antimicrobial hybrids (both functional elements of the hybrid express antibacterial activity) and antimicrobial conjugates (in which one functional element exerts its effect through non-direct antibacterial activity means). The non-direct activity expression may be through enabling active transport of the molecule as a whole into the bacterial cell structure or through interfering with the bacterial efflux process, while the other functional element expresses the direct antibacterial effect. Complications in this classification arise, however, on moving to potentially three or more different functional elements in the same molecule. It is considered better then to distinguish separate hybrid types as done in this book.

Designed hybrid compounds should be reasonably accessible synthetically with possibilities for scale up. Initially, though, smaller quantities are sufficient to assess properties (enzyme inhibition; cell culture) in vitro and to pursue initial mode of action studies including separate enzyme inhibition studies if required. Having scale up capability is important if progression to pre-clinical and then clinical studies are justified or indicated. Aside from these considerations, often there is also a complex interplay between chemical priorities and biological imperatives. A potentially fruitful area to explore in the synthetic design phase is that of using readily available, structurally advanced starting materials such as particular natural products or derivatives with known capabilities for interacting with biological targets. Synthetically, and to enable good SAR studies, one also needs to prioritise routes that are compact yet versatile, with tolerance of various functional groups, and with good stereoselectivity if required. Also one needs to allow scope for late stage changes in groups tailored for appropriate in vivo biological properties to meet drug absorption, distribution, metabolism, and excretion (ADME) and pharmacokinetic needs and with the flexibility to address possible toxicity issues. In the deliberate design of hybrid structures it is also important to embed scope for various structural modifications to improve potency.

3.1.2 Factors in Antibacterial Hybrid Design

Aside from synthesisability issues as mentioned above in Sect. 3.1.1, one can divide the factors for antibacterial hybrid design into those pertinent in vitro at the bacterial cell level and then additional factors to these which need to be considered for in vivo use at the pre-clinical and clinical level.

At the cell level in vitro, important factors include solubility of the hybrid to allow for addition to the cell culture at a range of concentrations, and to have the required degree of penetration or permeation depending on the positioning of the bacterial targets: on or near the outer cell surface; within the cell wall at different locations; on the cytosolic side of the cell wall; or intracellular. Also one needs to consider the likelihood or not of drug efflux.

Further factors require attention at the in vivo level and encompass oral vs intravenous administration; oral administration-stability to acidic pH, effect on the microbiome; ADME characteristics; other chemical transformations; then concentration at the bacterial site followed by the considerations delineated at the cell level in vitro. Clinical consequences after bacterial killing are also important but are not within the scope of this book.

In assessing ADME factors, ways to administer antibacterials, including unusual ones, are also important considerations and can have a major bearing on their effectiveness. Normally oral or intravenous administration methods are employed, while topical administration is used for bacterial skin infections, and suppositories can also be employed for other applications. For lung infections administration of the antibacterial through inhalation is an effective approach which bypasses exposure of the drug to the gut microflora and also high serum levels thus minimising off target effects. This pulmonary antibacterial delivery area and the associated difficulties with formulation has been well reviewed recently (Woods and Rahman 2018). For brain bacterial infections, the Blood Brain Barrier poses a significant drug delivery challenge, but perhaps nose-to-brain delivery might continue to be developed while taking into account other serious issues for such direct delivery (Schwarz and Merkel 2019).

A number of rules or guidelines have been developed to help with the design and development of new drugs including Lipinski's rule of five for oral bioavailability, structural features for oral administration and systemic absorption beyond the rule of five, and guidelines for permeation or penetration of Gram-negative pathogens. Aspects of the last two areas are discussed further in the following paragraphs.

From a retrospective analysis, DeGoey and co-workers (DeGoey et al. 2018) derived a relatively simple multiparametric scoring (MPS) function for orally bioavailable drugs beyond the rule of five as elaborated by Lipinsky. This function, formulated as AB-MPS = Abs(clogD-3) + NAR + NRB, correlated pre-clinical PK results of drugs and compounds in the collection of the American biopharmaceutical company AbbVie Inc.(AB), with Abs(cLogD-3), the number of aromatic rings (NAR), and the number of rotatable bonds (NRB) in the compound. Values of AB-MPS less than or equal to 14 were indicative of reasonable levels of oral absorption for compounds beyond the rule of five. Recognition that molecules are generally not rigid or inflexible entities is also important and further developments in this area cover this more explicitly looking at the effect of dynamically exposed polarity on solubility and permeability of chameleonic or adaptive drugs not compliant with the Rule of 5 (Sebastiano et al. 2018). Ways to quantify these chameleonic properties in macrocyclic drugs have been discussed by Whitty et al. (2016).

The recently developed structural guidelines for permeation of Gram-negative pathogens have been introduced in Chap. 1 (Sect. 1.3.3) and further detail is given here. These predictive compound accumulation or penetration rules in *Escherichia coli* were elaborated by the Hergenrother group (Parker et al. 2020; Richter et al. 2017). Surprisingly, compounds most likely to accumulate contain a sterically less-hindered (primary) amino group, are amphiphilic and rigid, and have low globularity (Richter et al. 2017). The Hergenrother group have also reported on the application

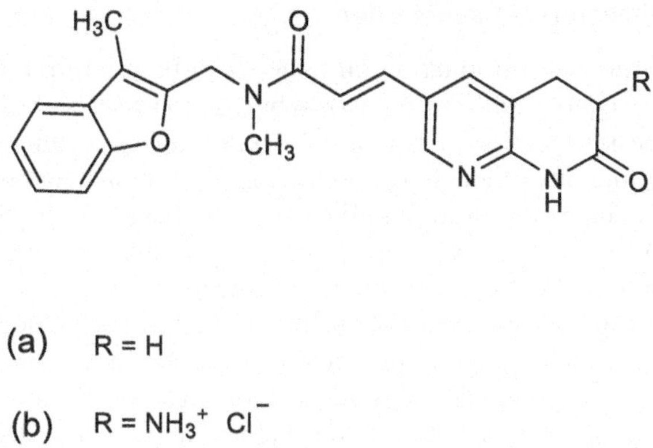

Fig. 3.1 Structures of the enoyl-acyl carrier reductase FabI inhibitors Debio-1452 (**a**) and Debio-1452-NH_3 (**b**)

of the rules via a web application they developed (eNTRyway; predicting compound accumulation in *Escherichia coli* from their strucures) and in association with structure–activity relationships and X-ray data, to the redesign of the enoyl-acyl carrier protein reductase FabI inhibitor Debio-1452 (Fig. 3.1a) to a derivative Debio-1452-NH_3 (Fig. 3.1b) which has good activity against a number of wild-type Gram-negative pathogens in vitro and in vivo with mouse models. Inhibitory activity against FabI was maintained with Debio-1452 –NH_3. The primary amino group positioning in this compound was decided from the eNTRyway-mediated prediction and from the results of computer based modelling and docking which indicated that enzyme target site binding would not be negatively impacted by the amino group at the 3-position in the reduced 8-azaquinolone ring system (Parker et al. 2020).

The incorporation of suitably positioned primary amino group functionality can be usefully considered as a late stage functional group transformation tactic as envisaged for example in the conversion of phenolic or hydroxyl groups to amino ether functionality. Specifically it could be useful in improving the penetration of berberine derivatives in Gram-negative bacteria through substituted derivatives incorporating an exposed primary aliphatic amino group in the substituent to help counteract the weak potency of berberine itself against Gram-negative bacteria.

An added dimension of complexity ensues when trying to treat bacteria residing within the host cell rather than being external to it. In the former case antibacterials then need to pass through the host cell membrane and avoid diversional interactions within that cell prior to bacterial uptake. Subcellular targeting has been reviewd by Rajendran et al. (2010) (see also Chap. 4, Sect. 4.2.3) and is based on the design of three component inhibitors with address and message structural components joined by a linker group but not specifically for antibacterials. It should be considered though for adoption in selective intracellular bacterial attack wherein the bacterium may be thought of as a 'host cell organelle'.

Structure—based design

Structural information on targets can be used in a number of areas to inform the design of multi-mechanism hybrid antibacterials. As for other medicinal agents, this design process is a rational one, if the target structures are known or can be modelled. Quite often though a combination of a rational and more intuitive design approach is employed, or an intuitive/analogue based approach. While rational target-based design has many positive aspects, it is also worth going beyond this and include intuitive or 'counter-intuitive' aspects of design alongside computer-based target design. Using both biologically as well as non-biologically-sourced structural toolboxes to express design ideas can also be very powerful resulting in what might be termed biological—biological, biological—non-biological, or non-biological—non-biological-type hybrids while tapping into a wider chemical structural space. Finding or confirming putative targets for the new hybrids or prodrug-derived compounds might also be assessed in the first instance using the Scifinder® database looking for similarities in structures between the proposed compounds and structures of compounds in the database with known bacterial targets or modes of action. Careful antibacterial testing would then need to follow to confirm activity modes or otherwise. Although in a different area, the structural similarity check aspect of this type of strategy has been used to successfully find presumed cellular pathways or receptors for hit compounds from a cell-based ROS assay, which subsequently identified the sigma-1-receptor as a potential new drug target for the treatment of the rare but devastating vanishing white matter disease (Atzmon et al. 2018). The Scifinder® database was used in this work for the similarity checking. Automated procedures are being developed to predict promiscuity which could be useful in complementing and refining the similarity searching (Hopkins 2009).

In addition, the automated design of ligands with polypharmacological profiles using adaptive design procedures is gaining ground and could be applied in principle to poly-active antibacterials where good structures for the bacterial targets are available. Besnard et al. (2012) have discussed this evolutionary approach starting from a clinically used anticholinesterase inhibitor (Donepezil) and evolving a range of ligands which could access the brain and display either specific polypharmacology or highly selective G-protein-coupled receptor features.

Fragment-based design

A knowledge of drug structures in combinations can be used to underpin at least the initial design of single multi-action agents. Simpler structural fragments containing the essential pharmacophoric units can then be identified and checked for activity prior to subsequent combining of the fragments and further testing. In this way potential multi-targeting hybrids can be accessed (Bolognesi 2013). The combination of two fragments may involve linking, fusion or merging processes. Linking is a common approach in general and one example of this in the non-antibacterial area is through use of an exocyclic double bond to link two separate pharmacophoric units (Gandini et al. 2018).

Fragment-based design can also be initiated by accessing starting fragments in the small to big approach, which can be a powerful one to develop new medicinal agents and has been used to develop antibacterial agents. In a neat example of this, a fragment-based approach combined with X-ray crystallography as a primary screen identified four hit compounds (from 352 fragment compounds) targeting the *Escherichia coli* bacterial sliding clamp, a protein vitally important in DNA replication. A tetrahydrocarbazole derivative then emerged as a good lead compound from further work and the methyl- and ethyl ester derivatives of this compound showed moderate antibacterial activity against some Gram-positive and Gram-negative bacteria (Yin et al. 2014). Further development of this intereting work has identified binding sites on the sliding clamp for tetrahydrocarbazole-based compounds (Yin et al. 2015).

In silico-based design

Various *in silico* approaches have been used to help in the design of multi-targeting ligands including those aimed at antibacterial activity (Ma and Chen 2012). Machine learning techiques have shown considerable potential in identifying new antibacterials (Ivanenkov et al. 2019b). Ivanenkov and co-workers started with a very large compound library (140,000 small compounds) with antibacterial activity against *Escherichia coli* and activity being assessed under the same conditions in the one assay. The compunds had a range of structures outside those of previously described antibacterials. Mining of this data then resulted ultimately in the identification of several compounds with potent activity in vitro and in vivo, although structural details were not revealed on these and very little on the mode of action apart from a mention of translation inhibition in prokaryotes and the induction response (to DNA damage) in some instances. Any multi-targeting mechanisms of action, while possible, await further assessment.

An in silico strategy was also employed in laying the basis for new drug designs for the treatment of the bacterial disease leprosy caused by *Mycobacterium leprae* or the more recently identified *Mycobacterium lepromatosis*. In this work the MUR enzymes (UDP-N-acetylmuramic acid or muramoyl containing ligases MurC, D, E and F) involved in the peptidoglycan pathway were targeted. Conserved or class-specific amino acid residues were identified, which then pointed to key interaction points to be taken into account in the design of any single molecule multi-action drugs (Anusuya and Natarajan 2012).

The powerful use of artificial intelligence (AI) and other computer-based methods for analysing multi-targeting in the context of antibacterial activity is presented in recent work by Abrusán and Marsh (2019). They based their work on ligands binding to sites across multiple protein chains. These sites were considered as likely to be conserved ones and thus suitable for targeting by broad-spectrum antibacterials. The multi-site binding ligands, which arose from the de novo ligand design plus deep learning, often contained, for example, sub-structural fragments with a degree of likeness to those in known antibacterials. Intriguingly, although perhaps understandably, this work revealed that multi-site binding also favoured compounds that were flexible although this came at the cost of predicted accumulation in Gram-negative bacteria through the increased number of rotatable bonds with increased flexibility.

Another interesting computer aided de novo general design approach to multi-targeting is described in work covering the LigBuilder V3 program (Yuan et al. 2020a), but with this approach the target structure is still required. However it is a promising program which can be used to design ligands to target multiple receptors or targets, multiple binding sites of the one receptor, or a number of conformations of one receptor. This program was employed to propose a new compound which should inhibit both HIV protease and HIV transcriptase at sub-micromolar activity levels. Application to bacterial target proteins of known structure should be feasible, and if confirmed, the best resulting designed ligands could also be assessed and further refined through the web application eNTRyway for likely accumulation in Gram-negative bacteria (*Escherichia coli*).

Natural products as starting points

As noted earlier, multi-targeting is the new paradigm in drug design and natural products form a good starting point as they are often promiscuous in their activity as noted by Chai and colleagues in their good review on the topic (Ho et al. 2018). Features of promiscuous compounds include commonly found pharmacophores and reactive functional groups, typically electrophilic moieties, and often such promiscuity is a problem as in the high throughput screening domain (Baell and Holloway 2010; Baell and Nissink 2018). The problem of nuisance sub-structures in drug leads also extends to some natural products (Baell 2016), but such compounds could still be useful as possible starting points for the ultimate development of a drug.

But how to design selective multi-targeting compounds which are not indiscriminately active is the key question here. In other words one would need to reduce the scope of the promiscuity while retaining controlled multi-targeting. There is a major problem though achieving selectivity for bacteria over host cells with this approach and ways to design-out unwanted off target interactions would need to be possible from modification of the core structure or functional groups in the natural product starting point, as in the case of berberine and analogues.

High throughput screening

Initial leads to mult-targeting antibacterials can also be accessed via high throughput screening (HTS) methodology based on large structurally-diverse compound libraries. While this approach may not be as appealing scientifically perhaps as more rational approaches, high throughput screening has resulted in some interesting new antibacterial leads representing new chemotypes. A good recent example of this was that reported by Ivanenkov et al. (2019a). They discovered a new 2-(pyrazol-1-yl)-thiazole–based chemotype using semi-automated HTS with a double-reporter system (pDualrep2) which identifies molecules capable of blocking the bacterial translational process or of inducing a response to DNA damage (SOS response). A large and diverse library of 125,000 compounds was screened. Some of these compounds displayed high activity against TolC deficient *Eschericia coli*; TolC is an important component of the AcrAB-TolC multidrug efflux pump in *Eschericia coli*.

Low to moderate cytotoxicity was noted with these pyrazolyl thiazoles (Ivanenkov et al. 2019a).

Other drugs as starting points

Drugs used for other disease treatments can also serve as templates for the eventual design of multi-action hybrids if they also show antibacterial activity alone or in combination. Such drugs can be found by a number of approaches including empirical screening or target based phenotypic screening or screening in whole-organism models (Farha and Brown 2019). A number of drugs used for other purposes have been identified with antibacterial activity as well (for example the anticystic fibrosis drug ivacaftor) (as referenced in Kaul et al. 2019). Drugs that synergise with antibacterials can also be identified through screening and such combinations would be of great interest to underpin novel hybrid designs. In related work, two existing drugs were repurposed successfully in a dual combination, antitrypanosomal (*Trypanosoma cruzi*) treatment as discussed in the review by Oldfield and Feng (2014). This review provides a good introduction to approaches to finding multi-targeting inhibitors as antibacterials.

3.2 Designing for Mainly Dual Activity

3.2.1 Dual Action Antibacterial Hybrids

Research work on multi-action hybrid compounds to date has largely focused on the design, synthesis and microbiological evaluation of dual action, or potentially dual action, compounds. These compounds have been extensively reviewed (Domalaon et al. 2018; Parkes and Yule 2016; Tevyashova et al. 2015; Shapiro 2013; Pokrovskaya and Baasov 2010; Bremner et al. 2007) so only some selected examples are given here. In general structural terms, many dual-mechanism or dual-action hybrids involve two molecular recognition components separated by a linker unit which may be a branched or unbranched chain or include the chain in ring systems. Omission of a linker then leads to other possible hybrid structural templates namely where the two molecular components containg the pharmacophores or target site recognition elements are fused sharing a common bond or joined at a spiro centre, and chimeras with partial to near full structural overlap. Note that the word ‘chimeric’ is derived from the chimera in Greek mythology, a hybrid creature which, interestingly, is usually depicted as a lion with the head of a goat arising from the body and with the tail being a snake with a head at the end i.e. a tri-component hybrid. In biology the term chimera refers to an organism whose cells arise from two or more zygotes and it is in this general sense the word has been transposed to the hybrid chemical structural framework.

With dual action hybrids, design is usually, though not always, based on results from the combination of two antibacterials with two different recognition elements

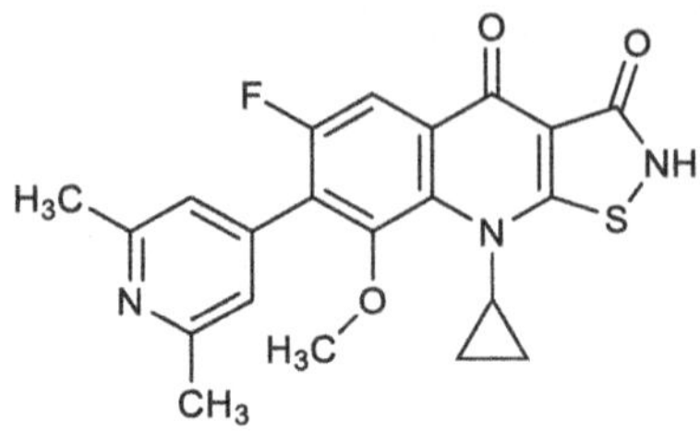

Fig. 3.2 Structure of a dual targeting isothiazoloquinolone derivative

A and B. In the single molecule hybrid the structural element A would be for a target A′ recognition site while B would be for the recognition of, or interaction with, a different bacterial site or target B′.

Alternatively, hybrid designs might include two separated pharmacophores which interact at two different sites on the one target macromolecule as exemplified by acrylamide–sulfisoxazole hybrid inhibitors of the bacterial enzyme dihydropteroate synthase in which the acrylamide moiety binds to the pterin binding site and the sulfisoxazole group interacts with the *p*-aminobenzoic acid binding site in this enzyme as noted by Nasr et al. (2020). Increased antibacterial potency was seen in these hybrids relative to standard antibiotics. Docking experiments were consistent with the proposed dual binding mode and *in silico* assessments of ADME properties were also undertaken which indicated that favourable properties could be obtained (Nasr et al. 2020). It would be difficult to extend this to more than two different binding interactions on the same target enzyme but not impossible and the strategy of interacting with three different sites on the one enzyme or other bacterially specific target should be further explored.

Dual targeting of both DNA gyrase and topoisomerase IV was demonstrated with an isothiazolone-fused quinolone derivative (Fig. 3.2) which showed low mutation frequencies at concentrations near the MIC values. These MIC values were in the low to very low sub-micromolar range against a wide range of *Staphylococcus aureus* strains (Cheng et al. 2007).

Another recently approved systemic antibacterial for community-acquired bacterial pneumonia, the pleuromutilin derivative Xenleta (Fig. 3.3), has a unique dual

Fig. 3.3 The chemical structure of the dual action systemic antibacterial Xenleta

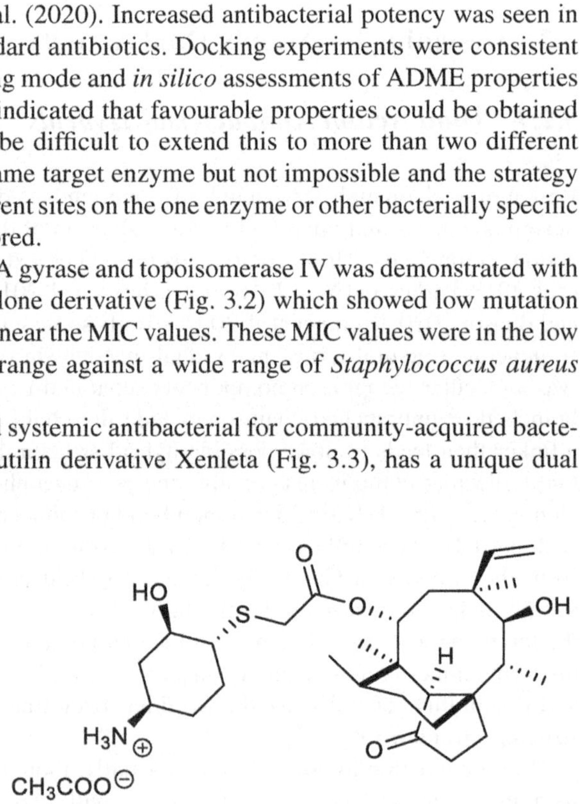

mode of action involving inhibition of protein synthesis through interactions with two sites, A and P, in the peptidyl transferase centre in the 23S ribosomal RNA of the 50S subunit. These interactions result in a closing of the bacterial ribosome binding pocket with negative consequences for the correct positioning of tRNA (CenterWatch 2019).

Novel binaphthyl-based dicationic tripeptidic derivatives have also been developed with good bactericidal activity against a range of Gram-positive pathogens. These compounds show very low resistance development properties in vitro and more than one mode of action has been tentatively proposed including cell-membrane disruption and the inhibition of cross-linking in the cell wall (Bremner et al. 2010a). The basic design rationale for these compounds centred on smaller and simpler cationic peptide derivatives with some related design features to the antibiotic vancomycin and which could still interact with the peptide-glycan moiety in the cell wall in both vancomycin-resistant and vancomycin-sensitive bacterial strains. Later work on cationic systems with a 1,2,3-triazolyl-containing substituent attached to the biaryl core produced potent antibacterials which depolarized the cytoplasmic membrane and permeabilized it in both *Staphylococcus aureus* and *Escherichia coli* (Tague et al. 2019b). Some related amphiphiles, which have both hydrophilic and hydrophobic moieties, also showed promising efficacy in vivo in a model of *Clostridium difficile* infection in the mouse (Tague et al. 2019a).

A new type of dual-mechanism of action antibiotic which incorporates activity against cell membrane integrity as well inhibition of folate biosynthesis via inhibition of dihydrofolate reductase (DHFR) has been reported in the pyrrolo-quinazoline SCH-79797 (Fig. 3.4a) and an in vivo active derivative Irresistin-16 (Fig. 3.4b). This combination of independent active sites resulted in broad spectrum bactericidal potency against both Gram-positive and Gram-negative bacteria with apparently no detectable resistance development (Martin et al. 2020). Membrane impairment assists drug entry and combining this type of membrane activity with one or two other antibacterial mechanisms of activity in the same molecule is a powerful strategy.

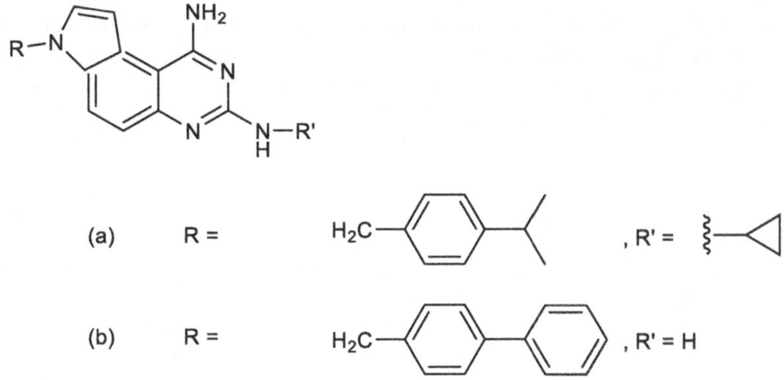

Fig. 3.4 Structures of the dual acting antibacterials SCH-79797 (**a**) and Irresistin-16 (**b**)

The results with the more potent Irresistin-16 suggested that the biphenylmethyl substituent is important for the membrane interaction and presumably not for the interaction with DHFR.

3.2.2 *Examples of Dual Action Agents from Nature*

Perhaps not surprisingly, bifunctional antibiotics have also been found in nature. One example is the antibiotic simocyclinone D8 (Fig. 3.5) (Edwards et al. 2009), which was isolated from the soil microorganism *Streptomyces antibioticus* Tü 6040. Simocyclinone D8 is only active against Gram-positive bacteria, and is cytostatic against some human tumour cell lines. Structurally it features a chlorinated aminocoumarin at one end and an angucyclic polyketide at the other, with a tetraene and deoxyhexose sugar unit linking the two. Simocyclinone D8 inhibits bacterial DNA gyrase by precluding DNA binding to the enzyme. From crystallographic analysis it has been shown that simocyclinone binds to the *Escherichia coli* gyrase A subunit via two binding pockets that separately interact with the polyketide and aminocoumarin structural components (Edwards et al. 2009). Both of these extra binding pockets differ from the inhibitory quinolone gyrase binding site, thus affording new opportunities for multiply active antibacterial design.

Another group of naturally occurring hybrid antibacterials are the thiomarinols A–G (thiomarinols A, C, D, E and F are shown in Fig. 3.6), which are metabolites of *Pseudoalteromonas* sp. SANK 73,390, a marine bacterium. These metabolites, of which thiomarinol A is the major component, incorporate two types of antibiotic structure: a functionalized polyketide-based acid linked via a fatty acid unit and ester and amide bond formation to a terminal dithiolopyrrolone moiety (Murphy et al. 2014). Thiomarinol A is a broad spectrum antibacterial and has structural similarities to the clinically used topical antibacterial, mupirocin, which lacks a terminal dithiolopyrrolone unit and has very good activity against the Gram-positive staphylococci and most streptococci. Mupirocin exerts its antibacterial potency through reversible inhibition of isoleucyl tRNA synthetase and hence bacterial protein and RNA synthesis. While the mechanism of action of thiomarinol A does not appear to be fully resolved one key aspect revolves around the dithiolopyrrolone unit which is activated by intracellular reduction of the S–S bond and then zinc ion chelation

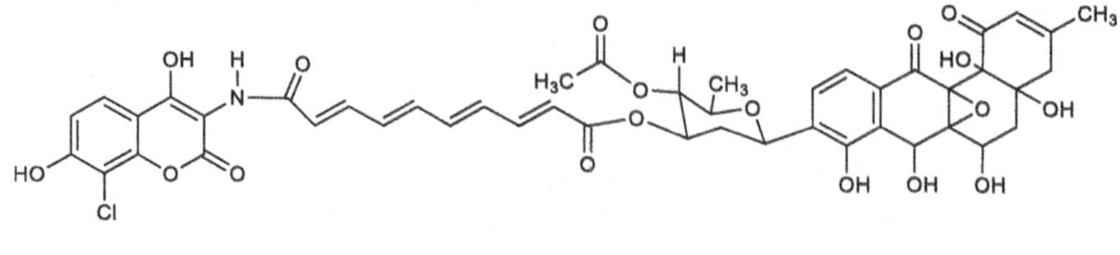

Fig. 3.5 Structure of the dual action antibiotic simocyclinone D8

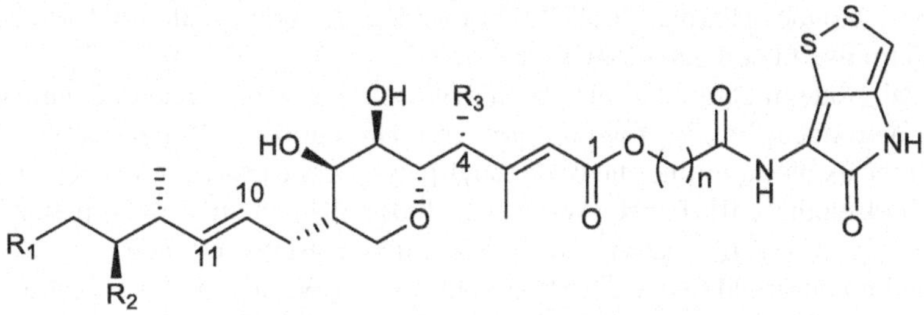

	R_1	R_2	R_3	n
Thiomarinol A	H	OH	OH	7
Thiomarinol C	H	OH	H	7
Thiomarinol D	Me	OH	H	7
Thiomarinol E	H	OH	OH	9
Thiomarinol F	H	=O	OH	7

Fig. 3.6 Structures of the antibacterial metabolites thiomarinol A and thiomarinols C–F

which then affects a number of metallo enzymes, including, significantly, metallo-β-lactamases in vitro (Chan et al. 2017; Li et al. 2014a). The dithiolopyrrolone unit is thus a prodrug-like group which remains attached to the rest of the molecule after activation by S–S bond cleavage.

The exciting discovery of teixobactin (Fig. 3.7) reported by the Lewis group in 2015 (Ling et al. 2015) and representing a new class of natural antibiotic with no detectable resistance development in vitro, triggered a major ongoing research effort in the area including analogue development. The polypeptide, teixobactin, inhibits cell wall synthesis by interaction mainly with the key conserved cell wall precursors lipid II as well as lipid III thus inhibiting the peptidoglycan and teichoic acid biosynthetic pathways. An additional mechanism for blocking cell wall synthesis has now been suggested involving the binding of precursors in clusters on the membrane (Shukla et al. 2020). Synthetic work in the area includes the development of a concise

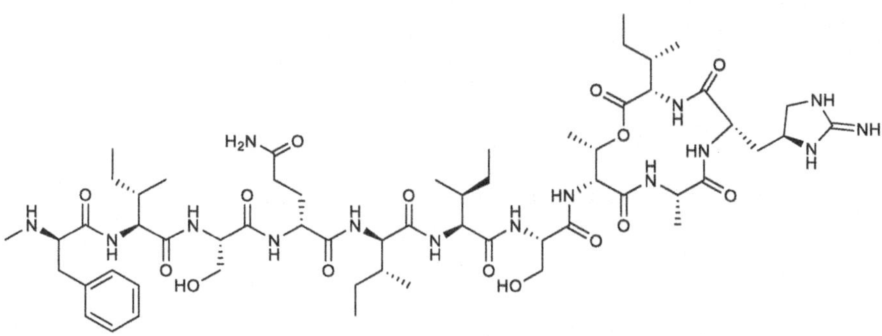

Fig. 3.7 Structure of teixobactin

route to analogues (Parmar et al. 2017) together the gram-scale total synthesis of teixobactin itself (Zong et al. 2019).

A dual mode of action has also been demonstrated with the natural antibacterial platencin produced by *Streptomyces platensis*. This antibacterial interferes with fatty acid synthesis by inhibiting the β-ketoacyl-[acyl carrier protein (ACP)] synthase II (FabF) and synthase III (FabH) (Wang et al. 2007). Plantencin displays potent broad-spectrum Gram-positive activity in vitro, including strains resistant to other antibiotics, and no observed toxicity from in vivo studies. While reduced susceptibility to resistance development might be anticipated through inhibiting two enzymes, caution is necessary as it appears possible in light of the self-resistance mechanisms observed in other strains of *S. platensis* which produce platencin and the related antibacterial platensimycin (Peterson et al. 2014).

Natural albomycins (Fig. 3.8) also have potent activity against Gram-positive pathogens and, interestingly, incorporate a trihydroxamic acid moiety as a siderophore linked to a terminal thiasugar-pyrimidine nucleoside unit and has been shown to be an inhibitor of seryl-tRNA synthetase. It is noteworthy that three hydroxamic units are involved to enable strong binding to Fe(III), presumably in order to compete with the natural bacterial siderophores (Lin et al. 2018).

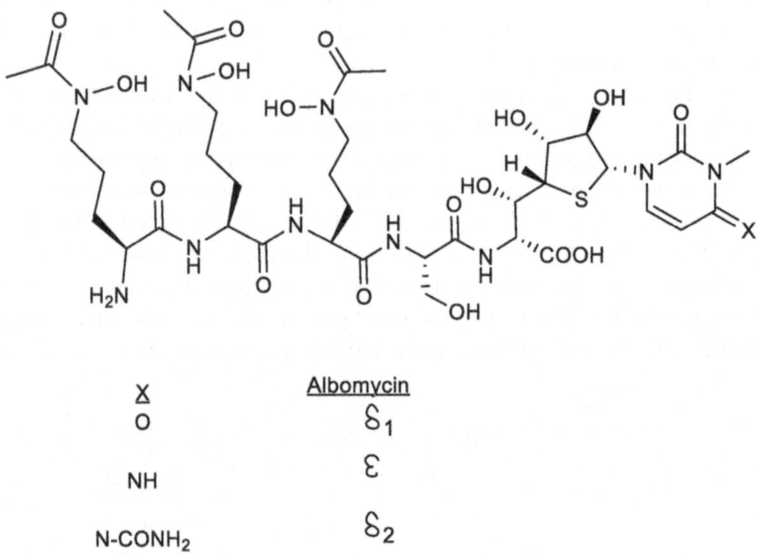

Fig. 3.8 Structures of the natural albomycin antibiotics (δ_1, δ_2 and ε)

3.2.3 *Berberine as a Starting Point for Design*

3.2.3.1 General Introduction to Berberine

Berberine is an isoquinoline alkaloid belonging to the protoberberine group and occurs in a number of plant species such as *Berberis* species in the Berberidaceae family. Berberine-containing plants are an important component in Chinese herbal medicines, and other herbal medicines, for the treatment of bacterial infections and for other diseases. Berberine is a quaternary salt which is readily available commercially and is usually used as its chloride salt (Fig. 3.9a, X = Cl). In this book the salt is simply referred to as berberine. It displays a wide range of biological activities and is a weak to moderate antibacterial with a number of modes of action. Berberine has been noted to have moderate *Mycobacterium tuberculosis* activity (Mishra et al. 2017). This protoberberine alkaloid provides a versatile structural platform for further chemical modification and reference will be made to it at other points in this book to illustrate this versatility, both actual and potential, for dual and higher activity design.

Berberine exerts a range of actions in biological systems through interaction with a number of targets including DNA and the FtsZ protein critical for bacterial cell division amongst others. The multi-actions of berberine are referred to in the article on

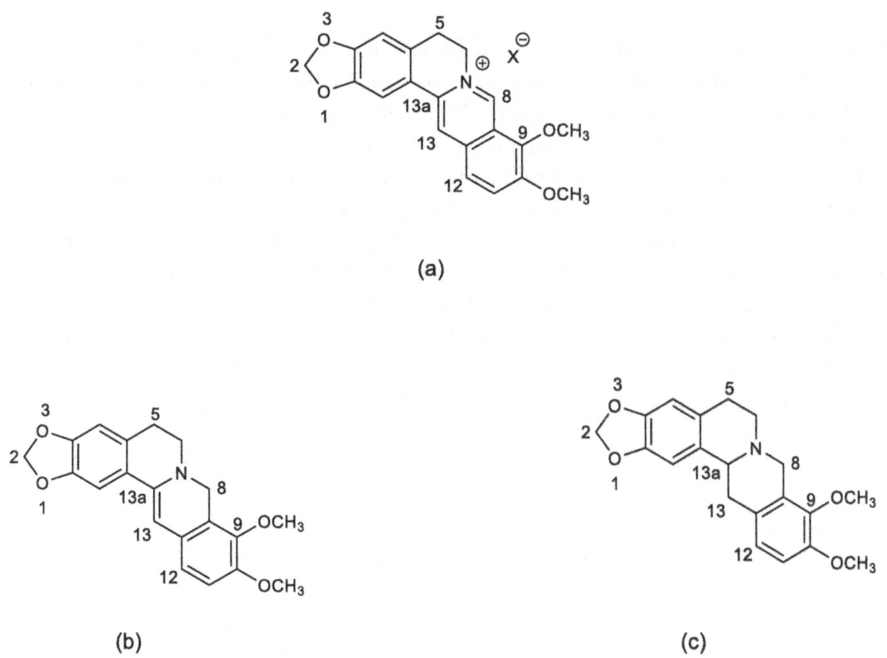

Fig. 3.9 Structures of berberine salt (**a**), dihydroberberine (**b**) and tetrahydroberberine (**c**)

the polypharmacology of natural poducts by Ho et al. (2018), and Wang et al. (2017b) have detailed the effects of berberine metabolism on its in vivo pharmacological profile. A summary of berberine metabolites, and berberine/berberrubine physico-chemical properties, plus the clinical use of berberine are also covered in the review by Caliceti et al. (2016).

As a result of its polypharmacology coupled with a versatile chemistry, research on berberine and derivatives continues to expand including in the antibacterial area. Implications for the design of potential multi-targeting agents based on berberine derivatives are explored further in this chapter.

It is also of interest to note that berberine has been shown to display anti-biofilm activity (Borges et al. 2015; Aswathanarayan and Vittal 2018). From in vitro, and in vivo (*Caenorhabditis elegans*), studies berberine inhibited biofilm formation (*Pseudomonas aeruginosa* and *Salmonella enterica* serovar Typhimurium) and had anti-infective properties with the *Salmonella* sp. at sub-MIC levels (Aswathanarayan and Vittal 2018). Also Sun et al. (2019) have shown that berberine at half its MIC concentration level can compromise biofilm formation through inhibition of the quorum sensing system in antibacterial-resistant *Escherichia coli*. This inhibition was effected through the downregulation of the expression of genes related to the quorum sensing system.

3.2.3.2 Versatile Chemistry of Berberine

Berberine has a rich and versatile chemistry which includes electrophilic substitution, nucleophilic addition and substituent group changes. Further versatility ensues after ready conversion to dihydroberberine (Fig. 3.9b) with its embedded enamine functionality and also further reduction to tetrahydroberberine (Fig. 3.9c). These reduced structures avoid issues which may flow from having a quaternary ammonium group present for instance with adequate oral absorption for in vivo activity, assuming the reduced structures have similar bioactivity. Dihydro derivatives are antibacterially active (Zhang et al. 2018), but dihydroberberine itself is not as active as berberine against *Stahylococcus aureus* in vitro (Rodrigues et al. 2018).

The **C8** position in berberine readily undergoes nucleophilic addition reactions with a range of carbon-centred and other nucleophiles (Nechepurenko et al. 2010). Examples include the formation of 8-acetonyldihydroberberine and 8-allyldihydroberberine by such nucleophilic additions. Reaction with potassium hydroxide and subsequent oxidation (presumably in the presence of air) affords 8-oxoberberine which can then be converted to the 8-chloro derivative by reaction with phosphorus oxychloride (Cheng et al. 2010). The 8-chloro derivative in turn is a very useful one for the introduction of other substituents for example 8,8-dialkyl groups (Cheng et al. 2010).

At the **C9** position the methoxy substituent can be demethylated simply by thermolysis of berberine chloride to give the 9-hydroxy derivative (berberrubine) in good yield. This phenolic group, although weakly nucleophilic due to significant electron delocalisation in the phenoxide ion, can then be functionalised further through *O*-alkylation reactions under basic conditions. *O*-Arylation reactions however required a pre-reduction of berberrubine with $NaBH_4$ to the terahydroderivative which could then be smoothly *O*-arylated via a copper-catalysed reaction with aryl iodides. Sudsequent re-oxidation with I_2/DMSO then gave the substituted quaternary ammonium iodide salt system in good yields (Teng et al. 2019). Chen and colleagues have also reported on additional *O*-alkylation and *O*-acylation reactions of berberrubine and tetrahydroberberrubine respectively to access further 9-substituted derivatives (Chen et al. 2017). *O*-Propargylation at the 9-hydroxy position also proceeds smoothly and the alkyne group then provides a useful handle for triazole-based linking of other moieties, increasing the versatility of this chemistry. For instance, this reaction has been used to access a range of such triazolyl derivatives substituted with various arylsulfonamide groups which displayed good antiplasmodial activity in vitro against *Plasmodium falciparum* (Batra et al. 2018).

The **C12** position in berberine is reactive towards electrophilic substitution such as bromination (Zhou et al. 2017) and nitration (Wang et al. 2020) providing direct access to groups which can also be elaborated further via selective synthetic methods either still at the ring quaternary ammonium oxidation level or at the tetrahydroberberine level and then later re-oxidation (Wang et al. 2020). Electrophilic substitution at C12 on berberrubine to afford a range of Mannich base derivatives in moderate to good yields has also been described (Li et al. 2014b; Mistry et al. 2017).

Extensive chemistry is also accessible through the **C13** position in the dihydroberberine derivative of berberine (Tang et al. 2017), or via 8-substituted dihydroberberines, as illustrated by C13 alkylation of the enamine moiety and subsequent re-formation of the berberine quaternary ammonium ion structure (Ball et al. 2006; Bremner and Kelso 2010; Park et al. 2006; Bhowmik et al. 2014). In one application of this, 8-allyldihydroberberine has been used in the enamine arylalkyation step followed by a [3,3]-sigmatropic rearrangement then a retro-ene reaction to give the 13-substituted berberine salt directly in a one pot thermal reaction sequence (Bremner and Kelso 2010). The enamine moiety in 8-acetonyldihydroberberine can also be hydroxylated with $KMnO_4$ and then converted on acid-catalysed elimination of acetone to 13-hydroxyberberine. The synthesis of 13-substituted derivatives of berberine, including 13-hydroxyberberine, is described by Tang et al. (2017). 13-Hydroxyberberine as the phenolbetaine is amenable to *O*-alkylation of the 13-oxy functionality (Samosorn et al. 2009) affording access to potential new derivatives incorporating other useful functional groups.

For other chemistry associated with 13-substituted berberines and some 8-oxo analogues, including transformation skeletally to the 7/5 series, see Zhou and Tong (2016). Other 13-substituted 8,13-dihydroberberines, still with a quaternary iminium nitrogen present in the ring, can also be readily accessed after enamine alkylation reactions. Fully reduced 13-substituted tetrahydroberberines can also be made facilely from these precursors by sodium borohydride reduction (Mari et al. 2020). The products in this case were investigated for their antiproliferative activity in vitro on NCI-H1975 lung cancer cells.

Incorporation of more than one substituent attached to the berberine core can be achieved by sequential application of the regioselective reactions mentioned above. A good recent review of a range of such berberine analogues, plus some interesting new chemistry for mono substituent variations, is summarised in Wang et al. (2015). Included in this review is a discussion of 13-alkyl- and various 9- or 13-alkoxy (long chain) berberine derivatives as well as disubstituted analogues, together with some antibacterial activity data. Such disubstituted berberines are readily accessible synthetically and some have reduced human fibroblast cell toxicity with very good in vitro activity against drug resistant strains of *Staphylococcus aureus* (Wang et al. 2017a).

In addition to the sites mentioned for the attachment of substituents, the fused methylenedioxy group in berberine can be selectively ring opened giving access to a 2,3-diol or catechol derivative (demethyleneberberine) on reaction with boron tribromide and aqueous workup (Roselli et al. 2016) and these groups in turn could then serve as anchor points for other groups, although regioselectivity issues may be a problem if different groups are involved. The methylenedioxy group can also be cleaved by hydrogenolysis to afford the *o*-methoxyphenolic system.

Apart from introducing other functional groups, the reaction products from cleavage of the methylenedioxy groups also constitute a manipulation of the berberine heterocyclic framework with the 5-membered ring being removed from the berberine 5, 6, 6, 6, 6 fused ring network. Framework manipulation of the berberine skeleton and reduced derivatives has been widely researched over a number of years and one example of this is the transformation skeletally to the 5, 6, 7, 5, 6 series as reported by Zhou and Tong (2016).

3.2.3.3 How to Devise New Structures

While de novo design or fragment-based design are important approaches to attaining multi-targeting compounds, an equally important opportunity involves starting with a known relatively complex structure (rather than simpler ones as in the fragment-based design approach) and using it to plan and develop new structures with different pharmacophoric groupings. To illustrate some of the principles involved with this alternative approach, application to the alkaloid berberine as a structurally advanced starting point is discussed in this sub-section.

The first general design steps normally involve maintaining the core framework and varying attached groups. Within the 'flatland' sphere of berberine there are sites or opportunities for substituent variation and for the introduction of other rings, including spiro systems, to increase topological complexity and other sites for displaying different functional groups with different relative spatial arrangements. A challenge though, and this applies to other medicinal agent design, is how to modify the skeleton expeditiously. Normally, to do this a synthesis from scratch is required, but with berberine as indicated in the section above, functional groups or group units can be introduced and/or interchanged with compact syntheses and moderate to good yields.

The second general design steps then usually focus on the core framework but keeping in mind the potential for new attached groups to be included. For example, topology and electrostatics can be modified by having the heteroatom in different ring sites and also introducing different atoms in the ring architecture. In embarking on this particular approach it is worthwhile initially assessing the maximum number of isomeric heterocyclic skeletons that may be possible to give some idea of the scope of the new chemical structural space which might be attainable. Also such systematic analysis can suggest possibilitities not previously considered and also enable directed literature searches to compare what is known with what is not. Aspects of this approach have been used previously for example in the non-fused and benz-fused oxaza and dioxaza-medium sized ring context (Bremner et al. 1982).

For berberine, the theoretical maximum number of ring isomers with one nitrogen in the 5, 6, 6, 6, 6 fusion of rings in the bent shape is 21 which includes replacement of oxygen in the fused 5-membered ring and of carbon at sp^3 positions, aromatic CH positions and quaternary carbon positions, some of which could also result in incorporation of a positively charged quaternary nitrogen centre. Including another nitrogen or a different type of ring heteroatom in this system would clearly enable a significant increase in the number of ring isomer possibilities. In terms of synthetic approaches to isomeric systems in general, it is also of value to then consider the general classification of approaches to heterocycles first elaborated by Stoodley and reported in 1977 (Stoodley 1977). Although published some years ago it is still very relevant as a starting point for synthetic planning in the organic and medicinal chemistry fields. Three general approaches, which also apply to carbocyclic synthesis, were formulated by Stoodley and include:

a. **Ring construction** where the number of rings are increased.
b. **Ring interconversion**, where the number of rings remain unchanged but the process may involve:
   (i) ring enlargements
   (ii) ring contractions, and
   (iii) ring relocations where neither ring enlargement nor ring contraction is involved but where a ring and a group of atoms react to give a new ring of the same size with a different arrangement of ring atoms. Heteroatom replacement of a carbon atom in the ring, or another ring atom, could be involved with no change in ring size or position, as for example in the conversion of phthalic anhydrides to phthalimides where O is replaced by N in the fused 5-membered ring.
c. **Ring destruction** where the number of rings are decreased.

The ring interconversion approach (b iii) involving ring relocation is of particular interest in the medicinal chemistry SAR context especially if it involves only a one point change. For example, relating this specifically to berberine, could an extra nitrogen be introduced adjacent to the extant nitrogen replacing carbon 8 or carbon 13a? No shape or little shape change would be expected but significant localised electrostatic potential change is likely. New hetroatom sites can also serve as different

Fig. 3.10 General representation of atom swapping in a ring by isomerisation

starting points for the controlled introduction of further substituents with the associated potential for the development of new chemistry. This general process is also referred to in the medicinal chemistry context as atom swapping (of one or more atoms with selectivity or precision), which is of great current interest in drug discovery and in a sense is like a chemical version of gene editing. If such selective atom swapping could be achieved predictably with precision and with a wide range of substrates at scale it would be revolutionary. Such atom swapping rightly forms a part of the proposed wish list of organic chemical reactions for drug discovery chemists (Krämer 2019).

Direct isomerisation is another way of achieving atom swapping and while Stoodley (1977) did not specifically refer to this process as such in his list of thirteen illustrative equations, it could be included as an added equation (Fig. 3.10) in his classification, where the lower case letters represent ring skeletal hetero-atoms and the transformation is an isomerisation with no change in ring size. These heteroatoms could be the same or different. Simplified versions would include systems with two or three hetero-atoms present with the transformation resulting in a change in the positions of these atoms relative to each other in the ring with a concomitant change in the way the hetero-atoms are linked. An example of this ring isomerisation process would be the photoisomerization of pyrazine to pyrimidine (Breda et al. 2006) where the number of linking atoms between the two ‘a’ atoms (nitrogens) changes but the 6-membered ring size is maintained. The representation in Fig. 3.10 applies when the heteroatoms are all different or when only two are the same heteroatom out of the four designated.

Furthermore it is important to consider ring relocation with a change in the ring skeleton but not ring size, in which the total number of rings is not changed in the operation with ring size retention. These relocation changes include ring shift/replacement/interchange and relocation with ring atom shift. Stoodley’s classification has also been incorporated and elaborated on in what is referred to as Diversity-Oriented Synthesis aimed at the synthesis of new potentially bio-active structural frameworks with the systemisation of build/couple/pair (ring construction) and ring distortion strategies. The latter strategy includes ring cleavage, ring expansion, ring fusion, ring rearrangement and ring-aromatization and combinations of these (Yi et al. 2018).

The basic Stoodley ring synthesis taxonomy can be extended to systems with more than one ring which in turn could highlight potentially under-developed approaches to these systems or suggest new approaches to target systems and thus aid in the exploration of further biological properties including antibacterial ones for multi-active antibacterial hybrid design. Such multi-ring systems could be bond linked. atom linked (spiro systems), bond fused, bond bridged, stable/isolable complexes with rings, quasi rings formed through non-covalent interactions, and mechanical bond linked systems such as catenanes, rotaxanes with more than one ring, pseudo-rotaxanes (Xue et al. 2015) and pseudo-catenanes. In the context of mechanical bond systems, it is of interest to note that nature uses antibacterial threaded-peptides like capistruin and probably, microcin J25, to inhibit bacterial RNA polymerase (Kuznedelov et al. 2011), and perhaps more such natural antibacterial systems will be found in the future.

It is profitable to explore this ring taxonomy theme further with berberine as the springboard for expansion into new chemical space while preserving capabilities for inclusion of pharmacophoric or functional group entities.

As noted in Sect. 3.2.3.2, 13-hydroxyberberine as the phenolbetaine is amenable to *O*-alkylation of the 13-oxy functionality (Samosorn et al. 2009) affording access to potential new derivatives incorporating terminal primary amino group functionality to facilitate permeation in Gram-negative bacterial pathogens. In addition, structural features to target inhibition of bacterial efflux pump activity could also be built in to an 8-alkoxy substituent leading to a hybrid which might overcome the pump efflux resistance mechanism while maintaining overall features similar to those in berberine and thus the ability to target the DNA and protein sites with which berberine interacts (Olleik et al. 2020). While 13-hydroxyberberine is a more potent antibacterial than berberine it has greater liver cell toxicity. However some other 13-arylalkoxy derivatives of berberine showed good antibacterial activity with very much reduced toxicity. Olleik et al. (2020) showed that these derivatives did not seem to fragment bacterial DNA but did interact with the bacterial FtsZ protein.

Thought experiments are important in helping to explore new structural space and coming up with new structural possibilities. With these experiments one imagines bond forming and/or bond making changes and then assesses the structural consequences of these changes. This can be illustrated for example with the reduced tetrahydroberberine core structure (shown for simplicity in Scheme 3.1 without the methoxy substituents or the fused methylenedioxy ring). Through such thought experiments one can visualise systematic bond breaking around the ring nitrogen as well as C–C bond breaking, followed by bond making or re-making, without necessarily knowing how to get there synthetically in the first instance. The emphasis should be initially on hypothetical structural changes. Later functional group manipulations or modifications with the addition or deletion of atoms and subsequent reactions can be incorporated leading to other skeletons. Commensurate with this, the simplified tetrahydroprotoberberine skeleton might be envisaged to provide access to benz-fused tropanes for example with substituent variations as illustrated in Scheme 3.1 by (I) and (II). This can be a powerful approach to new drug design in which envisioning experiments are used initially to arrive at the new structure, which can then be assessed for likely properties from database comparisons and from computed pharmacologically relevant physical properties. Then synthetic approaches

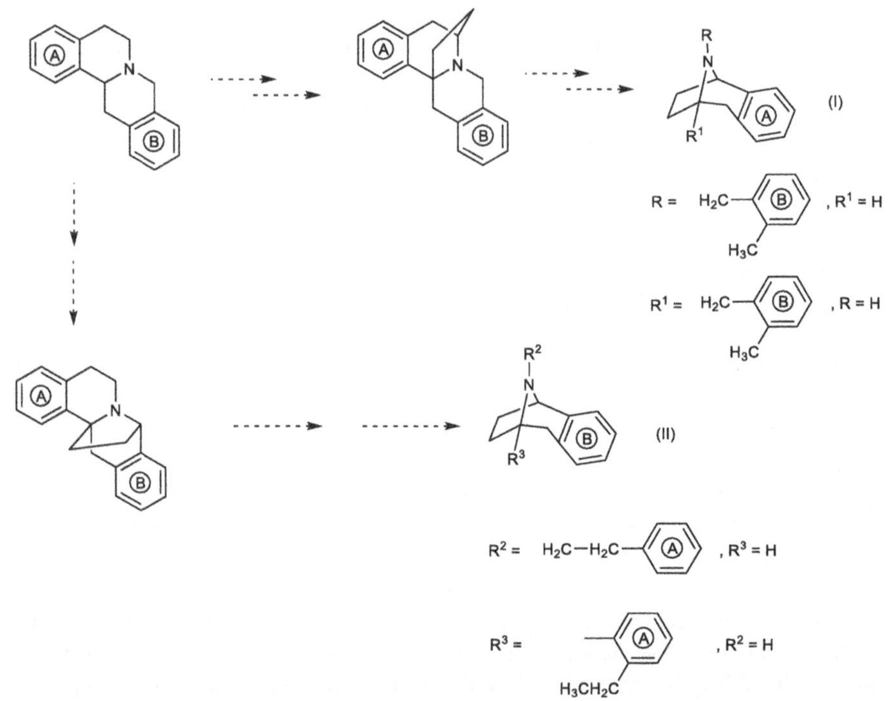

Scheme 3.1 Some illustrative 'thought experiment' transformations from the tetrahydroberberine-type core

would need to be assessed perhaps supported by computer-based (AI) proposed routes.

In the next steps after the design thought experiments it is important to then look at possible ways to synthesise the proposed molecular targets or move from design principles to design actuality or reality. While all possible synthetic ideas should be analysed at this point, just one potentially feasible route to the benz-fused tropane skeleton in Scheme 3.1 is described here. This system might be realised for example through application of the known 1,3-dipolar cycloaddition chemistry of 3-hydroxypyridinium salts on intermediate formation of the 3-oxypyridinium 1,3-dipole in situ (Lowe et al. 2020; Foley et al. 2017) (see also Sect. 3.3.4.1). A 13-hydroxyberberine salt, with an embedded 3-hydoxypyridinium salt type structural moiety, might reasonably serve as the substrate for the 1,3-dipolar cycloaddition with a vinyl sulfone derivative as the dipolarophile to access a bridged cycloadduct system (Fig. 3.11a) incorporating a benz-fused tropane nuclear core as highlighted in the alternative structural representation of this adduct (Fig. 3.11b). This addition product has different dispositions of aromatic rings, the tertiary nitrogen site, and carbonyl or sulfonyl-based functional groups as potential binding sites or sites for further transformations and the introduction of new groups. Quaternization of the

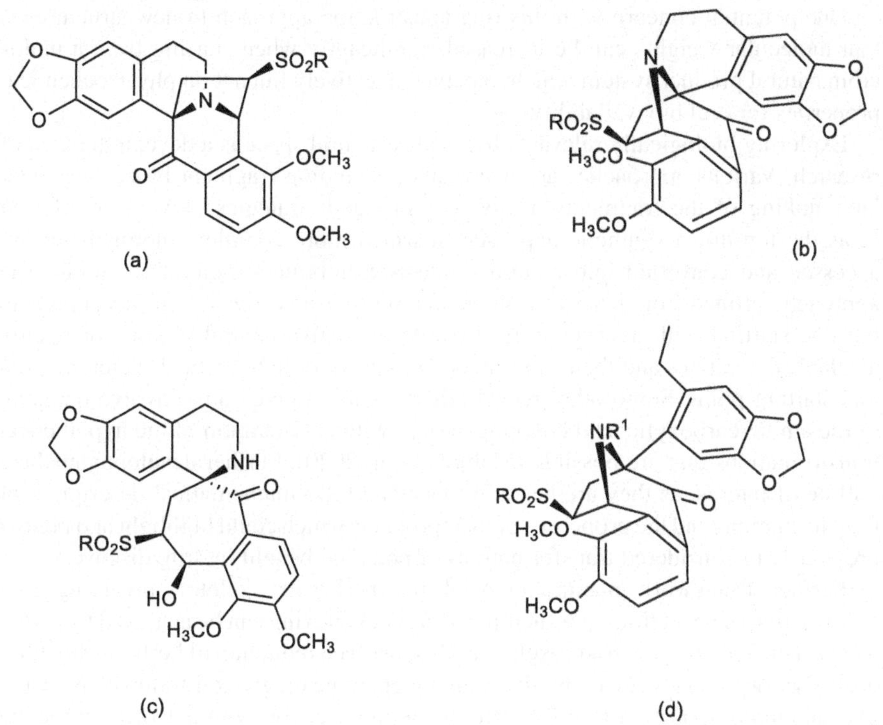

Fig. 3.11 Potential 1,3-dipolar cycloaddition product (**a**) from 13-hydroxyberberine, an alternative structural view of this adduct (**b**), and further possible transformation products (**c**) and (**d**)

tertiary amine in the cycloadduct, followed by Hofmann elimination and double bond hydrogenation, should then provide a system (Fig. 3.11d) with the core structure analogous to those in Scheme 3.1 (I and II) and with similarities in one of the substituents in one case (Scheme 3.1, (II), $R^3 = $ 2-ethylphenyl). The way one draws the 1,3-cycloaddition product can emphasise the original berberine core (Fig. 3.11a) or the embedded tropanoid core as shown in Fig. 3.11b. Changing the emphasis on one part of the skeleton over another can be very helpful in assisting with the synthetic design. This principle is worthy of general application in medicinal chemistry as it can also lead to new structural chemistry and provide a key stimulus to devise new reactions to meet a new need and/or use including in the complex multi-targeting paradigm.

The tertiary amine function in the resultant 1,3-cycloadduct (Fig. 3.11a) could also potentially serve as a starting point for transformation to the spiro system (Fig. 3.11c) via *N*-oxide formation, Meisenheimer rearrangement (Bremner et al. 1996) to an N–O bridged spiro system and reductive cleavage of the N–O bond to afford a new spiro rather than bridged system (Fig. 3.11d) with secondary amino and secondary hydroxyl group features as well as the presence of carbonyl and sulfone groups, and aromatic rings in different dispositions in three dimensional space.

One potential concern with this ring construction approach to new structures is that molecular weights can be increased significantly when starting from a multi-atom natural product system which may then negatively impact on physicochemical properties for oral bioavailability.

Exploring biologically relevant chemical structural space is a developing area of research. Various approaches are being taken including fragment based ones with later linking of the fragments in diversity-oriented strategies or what is referred to as the top down synthetic approach in which more complex intermediates are accessed and converted into natural product-resembling structural scaffolds. The approach outlined from berberine above in a sense is an extension of this approach, but one starting with a structurally versatile bio-active natural product or readily available derivatives and then exploring other known or hypothetical skeletons from this starting point. Some other recent advances in a related area involve mapping of the amine-carboxylic acid coupling system with enumeration of the hypothetical transformations that are possible (Mahjour et al. 2020). Further developments here will be of interest as they are heading towards a fully mathematical description of transformations and the properties of their products which could highlight and predict previously unconsidered transformations of potential benefit in drug discovery.

Further framework manipulation of the berberine skeleton involving ring construction of an additional 6-membered carbocyclic ring can be achieved by *ortho-* and *peri-*fusion on to the basic skeleton. This involved reduction of berberine to dihydroberberine and then attack by the resultant enamine on glyoxal followed by water loss and tautomerisation to reform the berberine skeleton with a 13-formylmethyl group. Cyclisation to give what is referred to as cycloberberine can then be achieved by intramolecular acid-catalysed electrophilic substitution on the adjacent electron-rich aromatic ring (Yang et al. 2019). Some cycloberberine derivatives with an 8-hydroxy group (the 9-position in berberine itself) show potent antibacterial activity due, at least in part, to the binding to the hydrophobic binding pocket in topoisomerase IV (Yang et al. 2019).

3.2.3.4 Dual Action Berberine-Based Hybrids

There has been a significant amount of work done on derivatives of berberine in a search for increased antibacterial activity compared with berberine (Nechepurenko et al. 2010; Iwasa et al. 1998) including in the dual action sector, where some intentionally dual-action designed berberine-based hybids have been made and assessed. Berberine is a substrate for the NorA pump which severely comprises its antibacterial activity, for example in *Staphylococcus aureus* so compounds have been made which contain structural elements based on a NorA efflux pump inhibitor attached to a berberine core. Some examples based on berberine (Fig. 3.12c) include 13-substituted derivatives incorporating methylene (Fig. 3.12a) (Bremner and Kelso 2010b; Tomkiewicz et al. 2010; Ball et al. 2006) or oxymethylene (Fig. 3.12b) (Samosorn et al. 2009) linker groups plus terminal structural elements of the potent NorA bacterial efflux pump inhibitor INF55 (5-nitro-2-phenyl-1*H*-indole). These

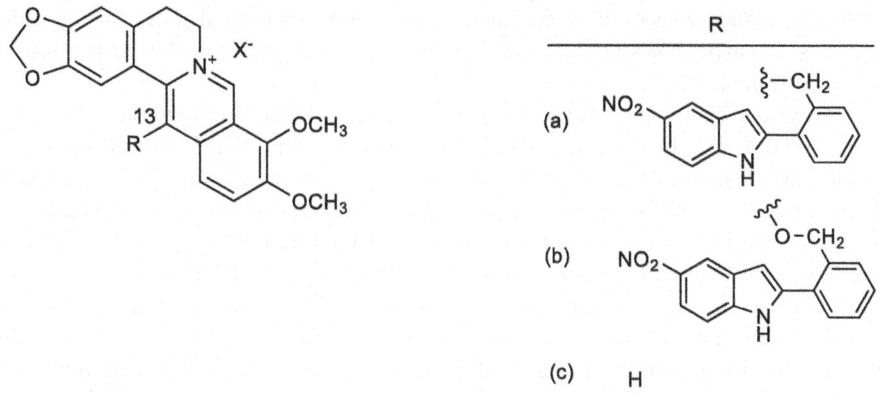

Fig. 3.12 Structures of some 13-substituted berberine hybrids with potential efflux pump inhibitor moieties

compounds showed very good activity against *Staphylococcus aureus* strains in vitro. Later studies on some *p*-linked methylene analogues related to (a) in Fig. 3.12 with N-methyl and NH indolic components indicated that the antibacterial mechanism or mechanisms of these particular berberine-INF55 hybrids were different from those seen with a combination of berberine and the INF55 type components (Dolla et al. 2014). This suggested these hybrids were acting differently from that envisaged in the initial design.

Inhibition (or evasion) of efflux pumps can be an important aspect of multi-action hybrid design including dual actions. There are many opportunities for small molecule binding with the range of different bacterial efflux pumps expressed and just one recent example of this is the binding of amidic molecules to the allosteric target protein AcrA, an essential component of the AcrAB-TolC multidrug efflux pump in *Escherichia coli* (Abdali et al. 2017). A fuller discussion of efflux pumps and their 'inhibitors', both competitive and non-competitive inhibitors, is covered in an informative review by Kapp et al. (2018). Another useful review looks at inhibitors and problems with clinical translation (Kourtesi et al. 2013).

3.3 Triple Action Antibacterial Hybrid Agents

3.3.1 General Points

There is considerable scope for the development of single molecule triple-action antibacterials. Synchronous or near synchronous action at three bacterial target sites could result in very potent antibacterial activity and greatly hinder resistance development. Some general design parameters for potential single molecule triple action antibacterials are discussed in this section, plus some specific known or suggested molecular expressions of these parameters. The extension of dual action design principles is a reasonable starting point for the design of non-cleavable triple action agents (Bremner 2017).

The general design parameters for triple action design are similar to those involved with antibacterial dual-acting hybrids (Tevyashova et al. 2015; Poäovskaya and Baasov 2010; Bremner et al. 2007) and prodrugs (Bremner et al. 2007), although there are extra possibilities when three separate interaction modes with respect to the biological targets are considered. Known dual acting hybrids as well as known combinations are good starting points for triple action design. In addition to direct antibacterial targets like those involved with cell wall formation, or bacterial DNA function or protein synthesis, the indirect activities could encompass variations such as dislocation of quorum sensing (Kalia 2013) and toxin blockade or the inhibition or attenuation of bacterial efflux pump activity.

Much has been done in the neuropharmacology area with regard to triple action anti-depressant design (Millan 2009) and this review is a helpful one generally for its good terminology discussion, and coverage of triple acting agents, higher order agents and the chemical challenge of the synthesis of such compounds. In the design protocol, the classes of ligand that were of interest in the clinical treatment of depression were considered first and then the single molecule hybridic design was based on exemplars in these classes. A similar approach could be adopted for single molecule multi-target antibacterials. In this context one could consider:

(i) Antibacterials that interact with a key target A and are more potent than other drugs targeting other sites.
(ii) Drug combinations that involve interactions at sites other than A.
(iii) Drugs with a clinically confirmed mechanism involving A and some other non-A mechanism.

In the antibacterial context not all ligands need be directly antibacterial, so one could look at a ligand which potently inhibits say bacterial DNA function plus two other ligands one of which blocks efflux pumps and the other blocks dihydrofolate reductase, and then try and incorporate the pharmacophoric elements in a single compound as discussed generally in the next Sect. (3.3.2).

3.3.2 Potential Design Based on Pharmacophoric Elements

In general terms one can consider firstly the pharmacophoric elements or recognition features as A, B and C for interaction with the bacterial targets. Features A, B, and C need not be those of the complete original drugs in each case but may incorporate similar characteristics to the original drugs. It is also important to try and minimize molecular weight increases if possible with this design approach. This can be illustrated in the following general design progressions starting with combinations and variations as the reference point and including three general pharmacophoric features or sites A, B and C as part of generic structures as indicated (Fig. 3.13). Variations

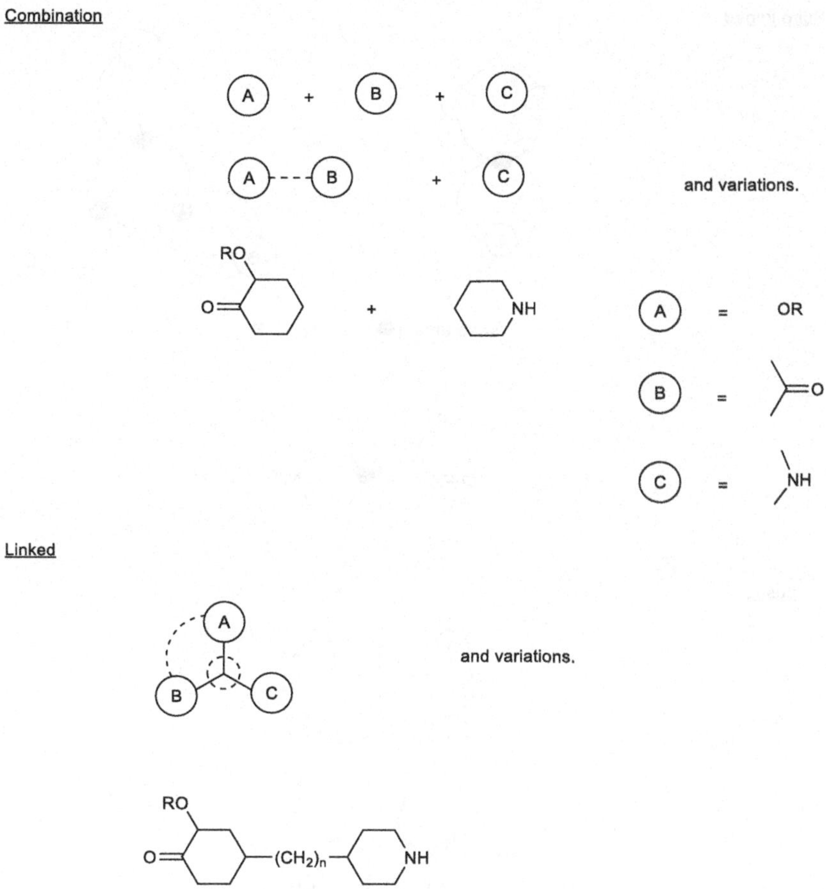

Fig. 3.13 General representations of the types of three interaction point hybrids derived from a combination starting point

are possible and with the three component starting point combination, the sites A, B and C would be part of three separate molecules respectively. In the two component combination, the dashed line between A and B in one of the components is meant to symbolize just one way of linking the pharmacophores amongst a number of variations which might include for example a linking ring system. Also included in the design progressions in Fig. 3.13 are some selected examples of molecular frameworks incorporating A, B and C as specified by an ether group (A), a carbonyl group (B) and a secondary amino group (C). Further detailed assessment of these classifications is given in Sect. 3.3.3.

The first general hybrid classification is the one involving linkage of A, B, and C in one molecular unit (Fig. 3.13). Again, the linkages could be incorporated in a variety of ways as indicated in the general structure. The next two classifications

Spiro joined

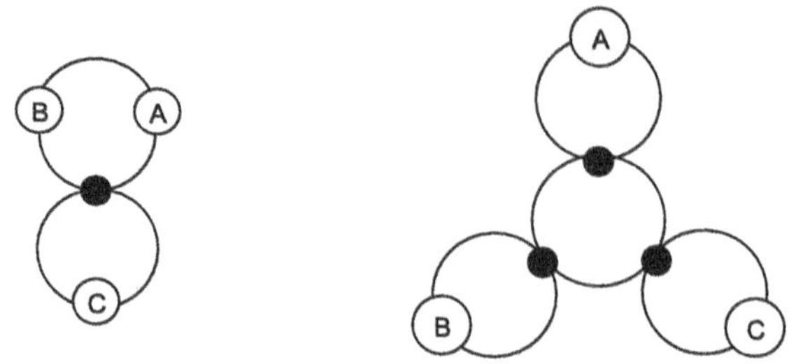

spiro atom (●) : C, Si, N^+, B^-

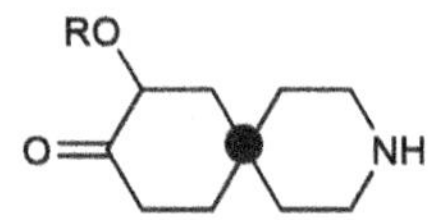

Fused

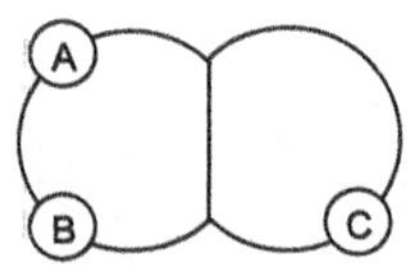

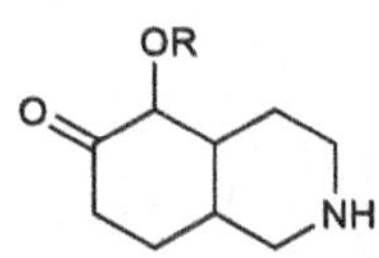

Overlapped
(Chimeric)

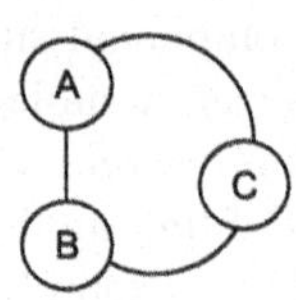

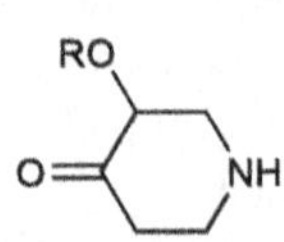

Fig. 3.13 (continued)

do not involve linking groups but the cyclic molecular units may be joined via a spiro centre or by sharing a common bond in a fused arrangement. The spiro linkage arrangement offers further interesting structural possibilities though incorporation of other elements in place of carbon at the spiro centre including nitogen, silicon, or boron each conferring different properties on the hybrid. With nitrogen it would carry a positive charge being quaternary, while with boron as a tetrahedral spiro atom it would have a formal negative charge. The use of silicon in place of carbon in potential drugs is an area of general interest (Ramesh and Reddy 2018) as are drugs incorporating tri- or tetravalent boron groups, and some boron-based antibacterials are discussed further in this book.

The last general classification (Fig. 3.13) involves an overlapping of the cyclic molecular units to give a chimeric structure.

In the representative models in Fig. 3.13, the 6-membered ring is maintained for inclusion of the pharmacophoric groups for clarity but clearly this ring size could be varied. In addition, extension of the models to include quasi 6-membered rings through strong non-covalent interactions would provide a powerful means to expand the skeletal design repertoire. For example with a 6,6-fused sytem the central 'bond' might not necessarily involve a classical covalent bond but might result from a transannular HN....C=O interaction involving the nitrogen lone pair electrons and the π^*–carbonyl group antibonding orbital.

Quasi-rings in antibacterial hybrid design

The use of quasi-rings resulting from non-covalent interactions to preference conformations in molecules has considerable scope in antibacterial design. Sometimes the term 'pseudo' (Greek) is used, but it is preferable to use 'quasi' (Latin) being more accurately reflective of the meaning of the word as 'resembling' or 'seemingly' in this context rather than 'false' for 'pseudo'. Such rings, include, for example, those formed from strong single (or more than one single) intramolecular non-covalent interaction(s) as replacements for actual covalent bonds while still containing other functionality. This functionality may either be attached to atoms at the non-covalent interaction sites or be located separately, for the required interactions with the biological target sites. Viewed from another perspective such interactions can be considered in terms of biasing conformational preferences and hence substituent spatial placements. A review by Beno et al. (2015) includes applications of this type of interaction in drug design with sulfur-based interactions.

Differences in geometries, atom separations and electrostatic potentials need to be considered in these interactions. Access to single molecule quasi-cyclic, quasi-spirocyclic, or quasi-bridged systems should be possible with the appropriate non-covalent interactions as well as combination systems which incorporate a molecular component with one or more quasi bonding motifs together with covalent bonds. The quasi-ring forming intramolecular non-covalent interactions can also be classified in terms of transannular (forming two quasi rings) and non-transannular or annular (quasi-single-ring forming) interactions. Generally these interactions result from electrostatic attraction or from some orbital interaction and partial electron distribution. A well established transannular interaction is evident with the alkaloid protopine, a modified isoquinoline derivative with some antibacterial activity, where the nitrogen lone pair is suitably positioned to interact with the carbonyl group carbon across the medium-sized ring in an n to π^* orbital mixing. High level DFT calculations confirm this interaction (Griffith and Bremner unpublished results). This interaction then constrains the conformation of the 10-membered ring to more closely match that in the protoberberine type alkaloid coptisine (two fused 6-membered rings). Coptisine is very closely related to berberine with a second methylenedioxy group present in place of the two methoxy substituents in berberine and it also has antibacterial activity.

In this example, above berberine was the starting point for the exploratory thought experiments, and proceeding from that point to the tetrahydroprotoberberine and then allocryptopine or protopine structural types. Calculations were largely based on protopine related to the berberine-type alkaloid coptisine as the rotational flexibility of the two methoxy groups in allocryptopine complicate the computations. These compounds themselves have weak and variable antibacterial activities but low cytotoxicities (Cheng et al. 2014).

There are many other possibilities for quasi-ring formation which might result in energetically favoured conformations with the overall bent shape of the berberine type system for further investigation of antibacterial properties and inclusion in multitargeting hybrids. One such hypothetical example involves a dibenzo-fused azecinone system (Fig. 3.14a). Such a system could incorporate a *trans* or *cis* configuration of the imine moiety and the possibility for a transannular interaction between the imine

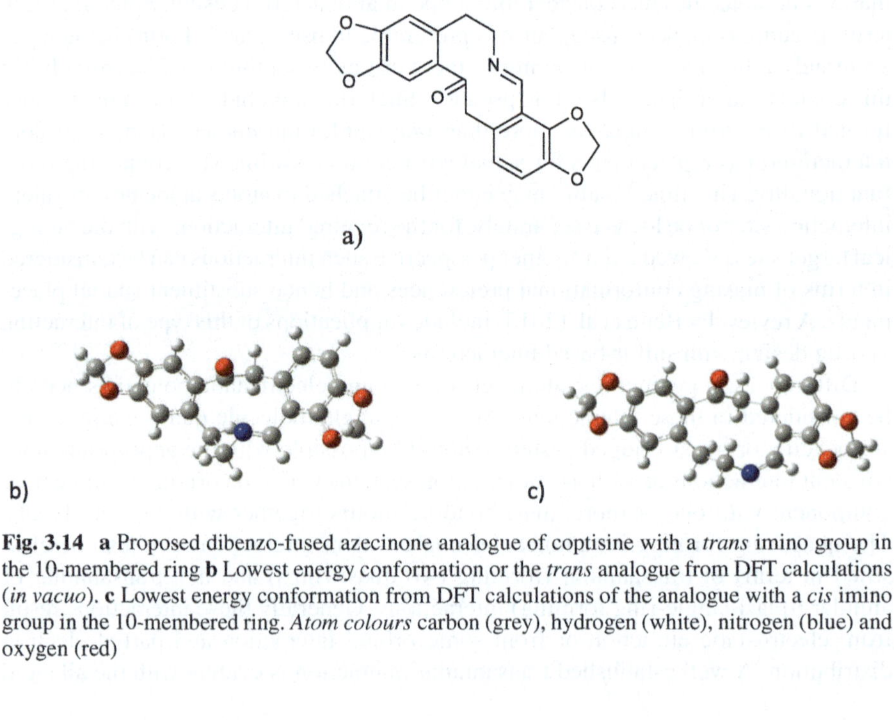

Fig. 3.14 **a** Proposed dibenzo-fused azecinone analogue of coptisine with a *trans* imino group in the 10-membered ring **b** Lowest energy conformation or the *trans* analogue from DFT calculations (*in vacuo*). **c** Lowest energy conformation from DFT calculations of the analogue with a *cis* imino group in the 10-membered ring. *Atom colours* carbon (grey), hydrogen (white), nitrogen (blue) and oxygen (red)

nitrogen and the carbonyl group embedded in the ring. This n to π* interaction was in fact indicated from DFT calculations to be accessible in the *trans* imine system in a favoured pinched conformation (Fig. 3.14b). In these high level DFT calculations, the structures were optimised *in vacuo* using the M062X functional and the 6–311++ g (2d, 2p) basis set. The pinched conformation for the *trans* imine was 3.61 kcal/mol more stable (*in vacuo*) than a wide conformation in which the imine nitrogen was well away from the carbonyl group. The optimised geometry conformation for the *cis* imine (Fig. 3.14c) was also a little higher in energy than the *trans* pinched conformer and no N–CO transannular interaction was observed in the former conformation (Griffith and Bremner unpublished results).

There are a number of potential advantages of quasi bonds over covalent bonds in the antibacterial design (and wider drug design) area. More compact or feasible syntheses of these non-covalent bond based ring systems may be apparent. Also they may open up possibilities for new chemistry and different synthetic routes. In addition, the restraint on otherwise rotatable bonds could give better control of *exo*-pharmacophoric group positioning or access to different positions and hence different binding sites on the biological targets. Such transannular or simply annular interactions also offer scope for the introduction of new skeletal atoms and new functional groups together with modified electrostatic potential energy distributions.

In addition, with non-bonded but interacting structural components, different or better pharmacokinetic characteristics in the molecule might be possible. These interactions would need to be maintained, however, in aqueous solution and on binding to the respective targets. Possibly, though, it may be advantageous for them to be overcome at the point of binding resulting in something like a quasi-prodrug situation and the intentional incorporation of such interactions in prodrug antibacterial design is worth considering further.

Annular non-bonded interactions have been shown to be important in the methylcarbapenem type antibiotics where an intramolecular non-covalent bonded S….$^{-}$OOC (classified as a 1,5-type with a quasi 5-membered ring formed) was indicated involving S in a thioether substituent and the carboxylate group O on the adjacent substituent (Nagao 2013; Nagao et al. 2001). Further annular non-bonded interactions of S with other atoms such as N or halogens have also been investigated by the Nagao group (Nagao 2013).

The annular interaction to form a quasi ring may also involve hydrogen bonding. An example of this from the anti-cancer drug area is the quasi-cyclization involving intramolecular hydrogen bonding of a pyrimidine substituted NH with a pyridine ring N to form a quasi-6 membered ring in new Mer kinase inhibitors (Zhang et al. 2013). This quasi ring replaced a pyrazole ring in a known Mer inhibitor (UNC1062) and further optimization produced a potent new series of inhibitors. Scope exists for further use of this quasi ring replacement strategy in the antibacterial intentional hybrid design area.

Another interesting aspect of non-covalent interactions is that they can also assist in, or drive, further covalent bond forming transformations as noted in the review of lactone macrolide—based antibacterials (Janas and Przybylski 2019).

3.3.3 Established and Potential Single Molecule Triple Action or Interaction Agents

One established example here is kanglemycin A, which is a naturally occurring ansamycin antibiotic related to rifampicin. Kanglemycin A (Fig. 3.15) mediates its activity through three interactions. Through its rifamycin-type core (A) it binds to a groove in RNA polymerase (like rifampicin) plus it binds just outside the groove (at a newly identified separate hydrophobic site) via a digitoxose residue (C) allowing it to inhibit the activity of rifampicin-resistant RNA polymerase. Kanglemycin A also has a structural extension involving a 2,2-dimethyl succinic acid-based unit (B) which inhibits the synthesis of RNA at an earlier stage than rifampicin via precluding the formation of initial dinucleotides. It shows good activity against rifampicin-resistant *Mycobacterium tuberculosis* (Mosaei et al. 2018). Peek et al. (2018) have also reported on the distinct mode of action of the kanglemycins compared to rifampicin and most recently (Peek et al. 2020) on some semi-synthetic analogues including a C3/C4 benzoxazino fused derivative with in vivo activity against MRSA and a rifamycin-resistant *Staphylococcus aureus* strain in a mouse model.

Triple action agents have also been developed from the potent antibacterial natural product vancomycin in elegant work described by Okano et al. (2017). One of their synthetic modifications included incorporation of a trimethyl ammonium group in an amide substituent attached at the C-terminal of vancomycin, plus conversion of a ring lactam moiety carbonyl group to a methylene unit in the binding pocket region, and thirdly derivatisation of the primary amino group in the terminal sugar residue to form a secondary (4-chlorobiphenyl)methylamino group. This resulted in a potent antibacterial which was much more potent than vancomycin itself against vancomycin-resistant Enterococci. Three independent synergistic mechanisms of

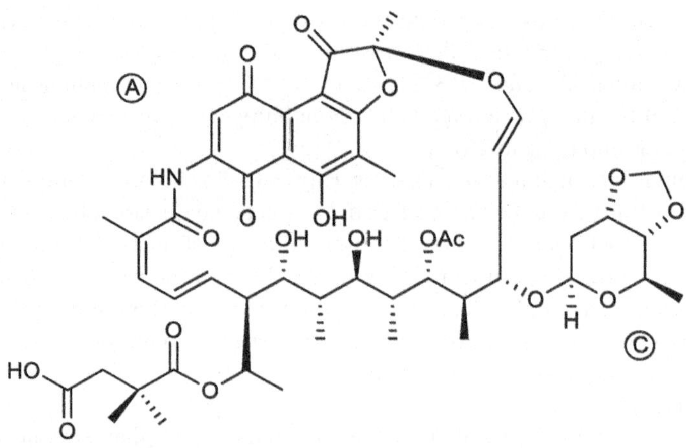

Fig. 3.15 Molecular structure of the ansamycin-type antibiotic kanglemycin A

action were displayed, one being induction of cell membrane permeability (from the quaternary amino group functionalized side chain), the second being inhibition of transpeptidase and cell wall biosynthesis (through dual D-Ala-D-Ala/D-Ala-D-Lac binding with the carbonyl group change), and the third being inhibition of transglycolase and thus cell wall biosynthesis (from the sugar amino group modification). Vancomycin analogues of this type showed a reduced susceptibility to acquired resistance by vancomycin-resistant Enterococci.

Other multi-action and potent vancomycin derivatives have been described and they involve functionalization at the vancomycin C-terminal through amide bond formation. One such group are the vancapticins, which have a linker group to an electrostatic peptide sequence (with positive charges for targeting and good interaction with the negatively charged phospholipid head groups in the bacterial lipid bilayer) and a terminal lipophilic unit to enable insertion in the lipid bilayer. In essence, one has a triple action mechanism operating with the third one being inhibition of transpeptidase and cell wall biosynthesis by the vancomycin core (Blaskovich et al. 2018a).

Further vancomycin conjugates with at least a dual action and possibly more have been described by Antonoplis et al. (2018). In these interesting and potent compounds compounds, vancomycin was conjugated via an amide linkage again (at the C-terminal carboxylic acid group and then a linker group) to D-octaarginine which served as a transporter moiety to sterilize MRSA biofilms and eliminate persister cells in vitro; this vancomycin conjugate also displayed excellent potency in vivo in a biofilm-associated MRSA mouse model.

Cumbre Pharmaceutics introduced a rifamycin-4-oxoquinolizine hybrid antibiotic (CBR-2092) for Gram-negative infections which acts by interacting with and inhibiting three targets: RNA polymerase, DNA gyrase and topoisomerase IV. CBR-2092 was then developed by TenNor Therapeutics as TNP-2092 (Ma and Lynch 2016) and has been granted FDA orphan drug status for the treatment of prosthetic joint infections. These infections can be difficult to treat due to the formation of bacterial biofilms. TNP-2092 (Fig. 3.16) has potent bactericidal activity against a range of pathogens that are associated with such biofilms (Fisher et al. 2020), together with a very good safety profile and a lower likelihood of resistance development as a result of the multi-targeting. In addition TNP-2092 shows very good antibacterial activity in vitro against clinical isolates of *Helicobacter pylori*, including strains with resistance to clarithromycin or levofloxacin (Ben et al. 2018). TNP-2092 also displayed good activity in a mouse infection model of *Clostridium difficile* and oral TNP-2092 showed some changes, considered to be positive, in the rat gut microbiome (Yuan et al. 2020b).

Compond TNP-2092 has a hydrazone unit linking the rifamycin and 4-oxoquinolizine portions. This compound displayed unforeseen synergy both in vitro and in vivo, compared to the separate drugs, but a query is whether it stays intact in vivo or undergoes hydrolysis. While hyrazones are generally more stable to hydrolysis than imines they are still susceptible to hydrolysis under acidic conditions and results suggestive of some hydrolysis is indicated in the work of Ben et al. (2018).

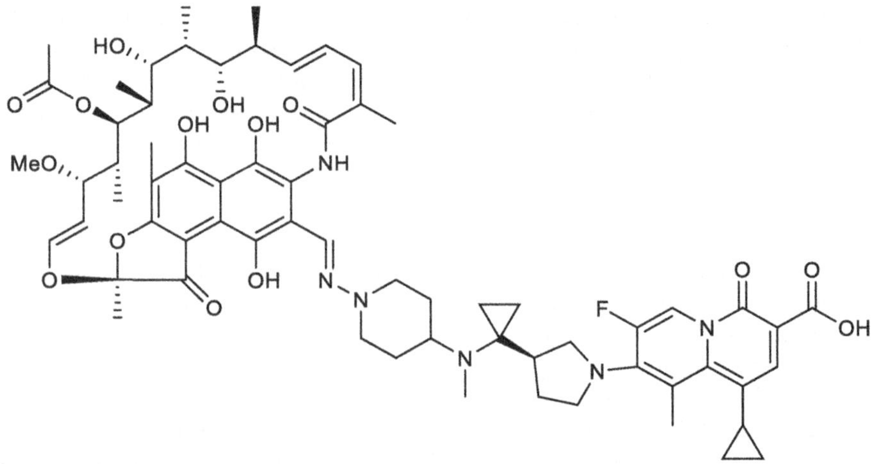

Fig. 3.16 Structure of the hybrid antibacterial TNP-2092

The linkage of tobramycin with moxifloxacin results in a hybrid (Domalaon et al. 2018) with the potential to interact with a number of bacterial sites including the 30S ribosomal sub-unit, as well as DNA gyrase, and through disruption of the outer membrane in *Pseudomonas aeruginosa* for example (Gorityala et al. 2016). Another hybrid including a fluoroquinolone is the potent ciprofloxacin-flavanone (naringenin) hybrid (Fig. 3.17) (Xiao et al. 2014). This hybrid binds to DNA gyrase with several additional interactions from the flavanone unit plus it either evades efflux or inhibits efflux pumps. The incorporation of quinolones or fluoroquinolones as valuable components in hybrid designs, together with the challenges involved, has been emphasised in two recent reviews by Gupta and Datta (2019) and Fedorowicz and Sączewski (2018), the latter paper being very extensive. A number of the hybrids discussed could conceivably have triple or higher binding interactions at one or more molecular targets.

It has also been indicated in other recent work (Tahir et al. 2019) on the design of ciprofloxacin-based hybrids that the group attached to the piperazine nitrogen

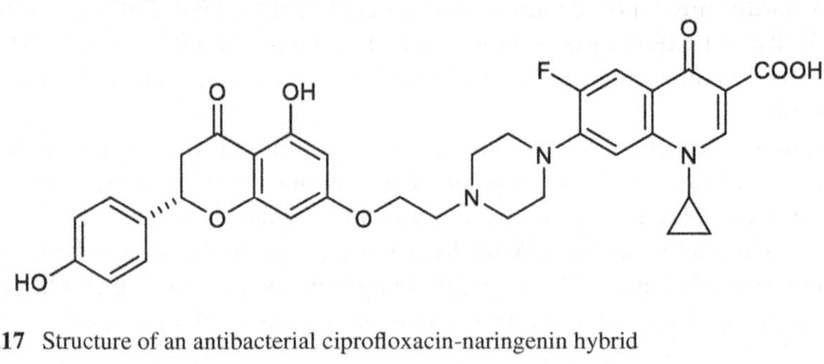

Fig. 3.17 Structure of an antibacterial ciprofloxacin-naringenin hybrid

could potentially impart a new mechanism of action by dual or multiple target interactions. Of the compounds assessed in this work, a ciprofloxacin-sulfanilamide hybrid (CGS-20) showed activity in vitro against Gram-negative bacterial strains, and Gram-positives, with MICs better than ciprofloxacin alone and comparable with the cephalosporin antibiotic, Cefixime. Dual or multiple targeting by the hybrid was not shown however (Tahir et al. 2019).

Siderophores have also been incorporated into triple action hybrids with the siderophore being a recognition element required for transport of the molecule into the bacterial cell. Such hybrids are based on the 'Trojan horse' approach. Nature uses this approach with the sideromycins (e.g., the albomycins), bacterial metabolites with a siderophore linked to an antibiotic moiety to attack other bacteria (Al Shaer et al. 2020). The linkage of siderophores to monobactam antibacterials offer considerable promise and not surprisingly this has been investigated significantly. One can envisage possibly a monobactam-siderophore compound with antibacterial and β-lactamase inhibitory activity as well (Decuyper et al. 2018), plus monobactams like aztreonam, which has excellent Gram-negative antibacterial activity through binding particularly to Penicillin Binding Proteins (PBPs) PBP3 and PBP1a, together with non-susceptibility to metallo–β-lactamases but not serine-β-lactamases, which can be co-expressed (Dean et al. 2018; Decuyper et al. 2018). The monobactam antibacterial, LYS228, may be better for inclusion in such a hybrid rather than aztreonam as it is stable to most serine-β-lactamases as well as metallo–β-lactamases (Dean et al. 2018). In research by Brown et al. (2013), pyridone-conjugated monobactam derivatives have been described with good antibacterial activity against clinically relevant Gram-negative bacteria. With these hybrids the hydroxypyridone pharmacophoric unit probably mediated binding to the siderophore receptors PiuA and PirA, while the monobactam unit would bind to a penicillin binding protein PBP3 (in *Pseudomonas aeruginosa*; Han et al. 2010).

In a potential extension of this approach, if the siderophore receptor also recognises and transports a pre-formed Ga(III) complex of the hydroxypyridone unit then perhaps this offers a way of selectively transporting Ga(III) intracellularly. If realised, one would have a mixed hybrid-prodrug approach in this case which would involve non-enzymatic release of the Ga(III) from the complex with the ligand still having PBP binding activity. While Fe (III) and Ga (III) are of similar size and ligand binding characteristics it may be possible to incorporate other ligands in the hybrid design to better complex Ga (III) while still being recognised by the transporter system as a substrate. It has been shown that Ga(III) has antibacterial and antibiofilm properties through actions affecting a number of biosynthetic pathways (Rzhepishevska et al. 2011). The potential of gallium complexes in therapeutic applications, including as antibacterials, has been reviewed by Lessa et al. (2012). Gallium (III) complexation in salicylidene acylhydrazides also could provide access to new multi-action antibacterials. The ligands themselves are of interest as anti-virulence agents which inhibit the type III secretion system in Gram-negative bacteria probably through multi-targeting in the cell (Hakobyan et al. 2014).

With a catechol-based siderophore attached to a cephalosporin, Cefiderocol (S-649266) has also shown potent activity against aerobic Gram-negative bacteria including multi-drug resistant Gram-negative pathogens (Ito et al. 2016; Aoki et al. 2018). The catechol moiety in this hybrid is an efficient iron chelator and the complex is actively transported into cells of *Pseudomonas aeruginosa* mediated by iron transporters. Binding of the hybrid mainly to penicillin binding protein 3 (PBP 3) but also to PBP 2 has been demonstrated (Ito et al. 2018). It can be speculated that a gallium (III) complex of Cefiderocol might also be worth investigation for ultimately delivering gallium (III) intracellularly with consequent negative flow-on effects for bacterial iron-based metabolic processes and biofilm formation (Górska et al. 2014).

Design principles for potential siderophore containing antibiotics, together with applications to mycobacteria, are discussed by Miller et al. (2009). In their article, Miller et al. also point out the siderophore-drug linker can be used for targeting and be cleavable so the molecule would then be a prodrug if inactive prior to cleavage. If non-cleavable it would be a hybrid.

A further extension of the Trojan horse design manifold might be to exploit the requirements for zinc (II) and manganese (II) by bacteria and develop zincophores and manganophores coupled to a dual action moiety such as ciprofloxacin, as long as this would result in synergistic actions. Manganese (II) uptake is mediated by two main types of membrane transport pathways involving ABC permeases and/or natural resistance associated macrophage protein (NRAMP) transporters (Eijkelkamp et al. 2015); Mn (II) can affect virulence of *Streptococci* (Eijkelkamp et al. 2015).

Another hybrid with triple activity is that derived from a neomycin structural template with a butirosin-based amide derivatisation of one of the amino groups in the cyclohexane unit and introduction of a double bond and deoxygenation of another sugar unit as observed in the antibiotic sisomycin. This chimeric trihybrid of elements from sisomycin (top right sugar unit) and butirosin (top left amidic side chain) around the neomycin core (Fig. 3.18) showed good activity against the ESKAPE pathogenic bacteria (Maianti and Hanessian 2016; Parkes and Yule 2016). The effectiveness of such a trihybrid of three antibacterial units augers well for the progression to other designs involving three antibacterials which could be based, for example, on the results from the evaluation protocols elaborated by Yeh and co-workers (Beppler et al. 2016) for combinations of three agents. Such combinations, and potentially others, are useful as a starting point in defining activity groupings in triple action agent design.

The trihybrid (Fig. 3.18) was potently antibacterial due to transport and accumulation in the bacterial cell, interaction with the ribosomal target possibly via a number of interaction sites plus the subsequent bactericidal action, and also avoidance or evasion of the major types of aminoglycoside modifying enzymes. The possibility of such evasion was an integral part of the design process for the trihybrid. This last outcome is not formally an 'action' but it highlights the need to include such considerations in the design of multiply active hybrids. In other words one can consider an approach of deliberate design for 'non-activity' expressed through eluding resistance enzymes (or efflux pumps) by inclusion of specific structural features, which can have a major impact in terms of antibacterial potency or spectrum of activity so it needs to be considered in the design context. Thus in the hybrid design process it is important to deliberately consider how to avoid certain targets and avoidance can

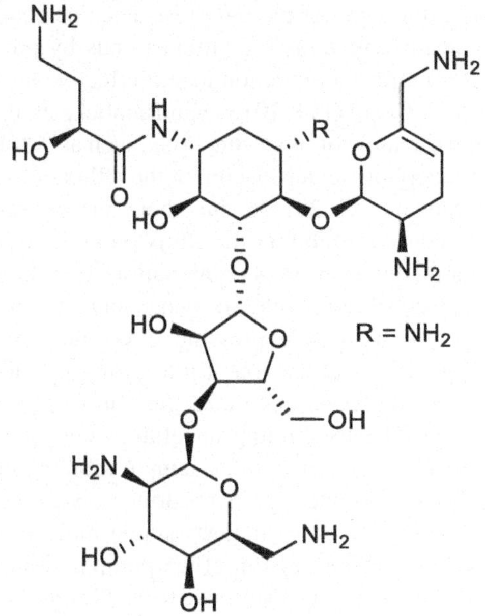

Fig. 3.18 Stucture of a trihybrid incorporating key components from the antibacterials sisomycin, butirosin and neomycin

be thought of as an 'action' by default. This has parallels with 'the curious incident of the dog in the night' which didn't bark as observed by Sherlock Holmes in the Sir Arthur Conan Doyle's story "The Adventure of Silver Blaze". This observation led to a different and productive line of thinking by Holmes resulting in a solution of the crime in this case. Similarly in multi-action hybrid design such considerations may apply.

Further examples of multi-active compounds include neamine heterodimers which might be at least triple acting as neamine can bind to two sites of the bacterial ribosome (Parkes and Yule 2016), and also the macrolide antibiotic solithromycin (CEM-101—developed by Cempra Inc.). Solithromycin, a hybrid derived from clarithromycin and azithromycin and the first fluoroketolide, has three interaction sites with the bacterial ribosome which greatly limits resistance development. It is a potent antibacterial against the most common CABP pathogens and was progessed into clinical trials. However in December 2016 the FDA rejected solithromycin due to concerns over liver toxicity (Buege et al. 2017; Owens 2017). Another example of two linked antibacterial units with more than two modes of activity are the hybrids described by Shavit et al. (2017) in which the aminoglycoside kanamycin is linked via a primary amino group in a glycoside unit to the secondary piperidinyl amino group in ciprofloxacin. The covalent linking incorporated two variable spacer units attached through a common 1,2,3-triazolyl moiety. While selected hybrids were poorer inhibitors than knanamycin itself of bacterial protein synthesis, they had similar inhibitory activity to ciprofloxacin against the target DNA gyrase and topoisomerase IV enzymes. Significantly, the hybrids also slowed down the development of resistance in *Escherichia coli* and *Bacillus subtilis*.

Although not the only resistance mechanism, bacterial efflux pumps are important for counteracting antibacterials by pumping them out of the cell resulting in concentration reduction to sub-effective levels in bacterial pathogens (Blanco et al. 2018; Costa et al. 2013; Van Bambeke et al. 2009). Additionally, efflux pumps also have a range of other functions, such as extrusion of endogenous metabolites like the siderophore, enterobactin, or the efflux of toxic endogenous metabolic intermediates (Blanco et al. 2016). Thus there has been a great deal of work done on the development of inhibitors for efflux pump inhibitors as a way to potentiate antibacterial action either through combinations or hybrids.

Incorporating other pharmacophoric elements in hybrid design for indirect activities with those expressing direct antibacterial activity can then be a powerful approach to efficacious triple acting hybrids. Such design can be based on inputs from drug combination results. In their review, Blanco et al. 2018 discuss antibiotic hybrids including an efflux pump inhibitor (EPI) which can synergise with other antibacterials, for example tobramycin-efflux pump inhibitor conjugates and fluoroquinolones against a drug resistant strain of *Pseudomonas aeruginosa.* This combination then suggests single molecule hybrid equivalents such as covalently linking the tobramycin-efflux pump inhibitor conjugate with a fluoroquinolone with the linkage being positioned to enable retention of the original modes of action. Hence such hybrids would be dual or higher order active agents which could form a good starting point for futher multiple-activity single molecule design. A simpler established version of this comes from the work of German et al. reviewed in Schindler et al. (2013) in which the oxafloxacin-efflux pump blocker (NorA) hybrid Q6CA showed highly potent EPI activity while retaining antibacterial effectiveness against *Staphylococcus aureus*, presumably with the latter due to gyrase and topoisomerase IV interactions giving a total of three activity sites.

Similarly, an antibacterial belonging to the benzoquinolizine fluoroquinolone group, levonadifloxacin (Fig. 3.19) and its oral amino acid ester prodrug (Wockhardt Discovery 2017) has demonstrated good bactericidal activity against quinolone-resistant and methicillin-resistant *Staphylococcus aureus* (MRSA). This compound targets the enzyme DNA gyrase along with topoisomerase IV. For the reinforcing third action, levonadifloxacin inhibits the NorA efflux pump in *Staphylococcus aureus* (Wockhardt Discovery 2017), which is another resistance mechanism deployed against fluoroquinolones together with resistance development through enzyme mutation.

Fig. 3.19 Structure of levonadifloxacin

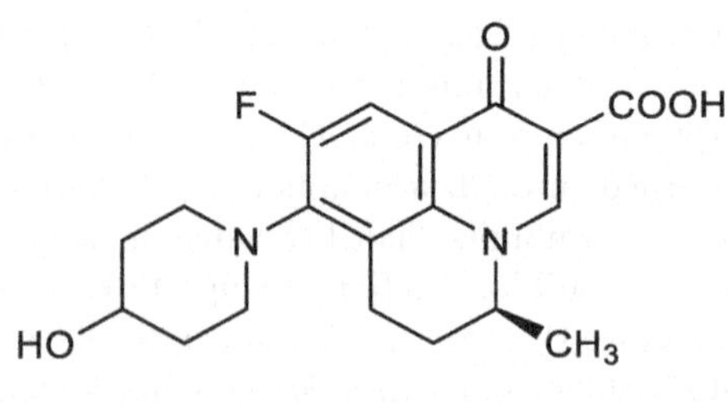

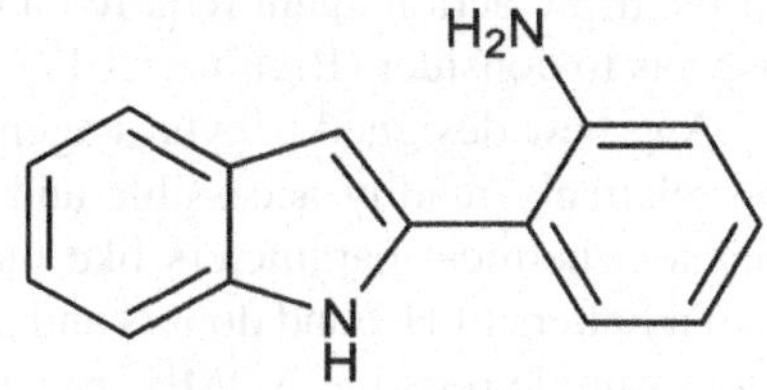

Fig. 3.20 Structure of the indolic efflux pump inhibitor RP2

Inhibitors of Gram-negative efflux pumps are of great interest and an earlier review by Opperman and Nguyen (2015) provides useful detail on the molecular basis for the inhibition of Gram-negative efflux pump-RND pumps. Inhibiting efflux pumps can also have other inherently deleterious bacterial effects and inhibitors of efflux pumps have been shown to slow the emergence of resistance as well as to reduce biofilm formation and bacterial virulence (Blanco et al. 2016). Another comprehensive review by Lamut et al. (2019) covers efflux pump inhibitors of clinically relevant multidrug resistant Gram-positive and Gram-negative bacteria (Lamut et al. 2019). Efflux pump inhibitors of Gram-negative bacteria are examined in reviews by Amaral et al. (2014); Blanco et al. (2018), and and more generally in a range of bacteria by Mahmood et al. (2016).

In the design of new multi-active hybrids which incorporate an efflux pump inhibitor pharmacophore use can also be made of preliminary *in silico* screening and docking on a modelled transporter as done with respect to the NorA transporter as described by Tambat et al. (2019). This paper also reports the isolation of a potent efflux pump inhibitor 2-(2-aminophenyl)indole, RP2 (Fig. 3.20) from a terrestrial *Streptomyces* sp. IMTB 2501. Evidence that RP2 inhibited the Nor A pump as well as the TetK and MsrA efflux pumps in *Staphylococcus aureus* was obtained from synergy studies in combination with other antibiotics in this bacterium. For example, RP2 potentiated the activity of ciprofloxacin in *Staphylococcus aureus* both in vitro and in vivo in a mouse thigh model infection study. The structure of RP2 has interesting structural similarities to the potent synthetic NorA pump inhibitor INF 55 (5-nitro-2-phenylindole), and analogues (Ambrus et al. 2008).

3.3.4 Designing Potential New Non-cleavable Triple Action Agents

Numerous combinations and permutations are feasible within a single molecule with respect to key recognition entities and their bacterial target sites. While these possibilities can be useful, synthetic challenges as well as efficacy challenges usually result, particularly as binding at one target, for example by molecular recognition unit A, may be negatively impacted—in steric or electronic terms or both-by the near presence of B and/or C. The possible effect of the other recognition units and linker groups on interaction with each target site, as well as the effective concentration of the triple action agent required and achievable at each site, are important further aspects to consider (Bremner 2017).

Any new designed trihybrid agents should be likely to be chemically stable and be relatively readily accessible and scalable synthetically. In addition, a range of physicochemical parameters like molecular weight, water solubility, lipophilicity, and numbers of H-bond donors and acceptors need to be considered, amongst others (for example possible ADME properties) for potential oral administration, although intravenous administration is an alternative if this is not feasible. Physicochemical parameters for any new proposed hybrids could be calculated using appropriate software like the Molecular Operating Environment software (MOE) as done in the case of an analysis of 14 different physicochemical features of known anti-tuberculosis drugs (Koul et al. 2011). In new drug design potentially toxic structural moieties also need to be avoided together with functionality likely to be problematic metabolically. Biologically, the potential antibacterial activity spectrum (particularly for Gram-negative pathogens), potency (particularly against drug resistant strains), the assessment of resistance development to the new agents, and possible off target effects all need to be taken into account.

Further pre-synthesis assessment of proposed new agents can be undertaken using the open access program LLAMA for drug-lead likeness and molecular analysis before deciding on synthesis (Colomer et al. 2016). The drug lead likeness alogarithm involves determining what is termed a 'lead-likeness penalty' for each structure rather than applying strong filters. The molecular analysis capability in LLAMA can also assses the originality of the molecular structure proposed.

After the initial design proposal it is always useful to look at refinements based on *in silico* considerations and literature analysis. It is also suggested that it is beneficial to look as well for ways to test or partly validate designs for triple activity in a preliminary way if biological target structures are known and other ligand interactions with the target are known with some precision. That is, try and go one step beyond intuitive structural suggestions.

In molecular terms, intentional non-cleavable triple action designs can be grouped in many different ways, but in this book the non-cleavable designs have been sub-divided into six types (i. to vi.) (Sect. 3.3.4.1) with structural recognition elements generalised for simplicity by the letters A, B, and C. These elements could involve functional groups embedded in, or attached to, one or more rings or acyclic units.

3.3.4.1 General Non-cleavable Types with Three Different Recognition Elements

In the design types below, dashed lines are used to indicate variations in the nature of the linking groups between the recognition elements indicative of the many different linkage architectures possible. In turn this affords much scope for different relative dispositions of the three different recognition element regions as designated by A, B, and C. The recognition elements represent the structural patterns likely to interact with the specific bacterial target sites. As these are general design motifs, in each case other possible variations with different ordering of A, B and C is assumed and where dashed lines are used to indicate the possibility of a range of different structural categories linking the recognition elements. For all these structural types, the design needs to be such as to resist chemically- or enzymatically-based cleavage so the compounds reach their target biological interaction sites intact.

Many structural design expressions are incorporated within each classification type but by doing this it can be helpful in systematizing the structural variety and highlighting gaps and areas for further development. For clarity the beginning of each type is highlighted in bold together with the specification of the sub-section.

3.3.4.2 Type I. A---B---C

The two linking groups as represented by the dashed lines which characterize this type i motif could include acyclic (linear or branched) groups with one or more atoms and/or one or more cyclic groups having spirocyclic motifs. With this general type i, the whole assemblage, or parts thereof, could also be intrinsic to one or more of the ring systems, with up to three linking groups or units present. One expression of this, for example, could involve A being joined to C by another linking group. Incorporating such rings in the linkers offers the prospect of defined and different relative spatial distributions of the A, B and C recognition patterns.

Further variation in the spirocyclic template architecture can be obtained through a stereogenic element at the spiro centre which can also involve both carbon or other atoms like a quaternary nitrogen. Additional expressions of spirocycles in triple action hybrid design could include C, Si and B as the spiro atom or atoms thus giving rise to neutral (C, Si) or charged (negative—B^- or positive—quaternary N^+) sites together with the appropriate counterions. Also one could include the possibility of betaine systems (B^-, quaternary N^+) in the spiro structural motif. The relative rigidity of the rings can be controlled by ring size. The pharmacophoric groups could be incorporated as substituents on (like a hydroxyl or amino group) or in the rings (like a carbonyl group). To help with the spirocyclic design process, useful parameters for desirable physicochemical properties and the molecular shape index in the context of some spiro-heterocyclic systems have been outlined by King et al. (2019).

Spirocyclic systems are an increasingly important part of modern drug discovery (Zheng et al. 2014), including antibacterials as exemplified by the gyrase B inhibitor, Zoliflodacin, which is in Phase 3 clinical trials for the treatment of infections caused by drug resistant *Neisseria gonorrheae* (Bradford et al. 2020). Zoliflodacin contains a spiropyrimidinetrione unit with carbon as the spirocentre atom. This unit has linked lactam and imide features in the pyrimidine ring. In general design, the embedding of a secondary lactam moiety in one or both of the spirocyclic rings is worth exploring further as it offers other opportunities for target site interactions through hydrogen bonding rather than the spirocycle being just a carbocyclic scaffold. Separating the NH and carbonyl groups of the original lactam by the spirocentre carbon would also give different opportunities for such interactions or for the attachment of other groups with their associated recognition elements. Advantages that might accrue from spiropeptidic designs include positioning control of functionality and resistance to enzymatic attack on the core plus compactness, rigidity or partial flexibility, and different chiral elements. Many spirocyclic antibacterials are known in nature, one example being pseurotin A from *Aspergillus fumigatus*, which has some moderate antibacterial activity against Gram-negative and Gram-positive bacteria. Pseurotin A has a chiral partially peptidic 5,5-spiro system with an oxygen heteratom in one ring in place of an NH (Mehedi et al. 2010).

In the hybrid design process it is good to have a basic starting skeletal structure with versatile functional groups incorporating say three pharmacophoric elements. In this book where examples are suggested, normally, although not only, two basic skeletal types are used, one being based on the tropanoid skeleton initially and the other on the berberine skeleton and related variations. In essence this will illustrate approaches starting from a more globular sp^3-atom rich structure in the first case, and a flatter structure with aromatic fused rings in the second.

With the tropane derivatives, many opportunities present for the ready incorporation of three different pharmacophoric units in substituents attached to bicyclic and tricyclic templates with a high proportion of saturated carbon atoms.

On the other hand, with the berberine-based systems, one has access to flatter molecules. While a common sentiment in drug design is to move away from flatter structures to improve the possibilities for clinical success, flatland can still be interesting both chemically and pharmacologically but structural hills and valleys should also be explored. In an interesting report by Monteleone et al. (2017) an analysis was done on how molecular connectivity can predefine polypharmacology or not, looking at the classification of active molecules although not specifically antibacterials. Drugs with a higher number of double bonds and fused aromatic rings tended to be active on more than four targets while one target selectivity was more evident with higher numbers of sp^3 carbons and aliphatic rings. This suggests for multi-active hybrid design that the possibility for incorporation of more tetrahedral carbon sites in flatter molecules might still be a productive line of study. Higher levels of skeletal sp^3-atoms do allow for a greater diversity of sites in chemical space to be explored as noted in Morgentin et al. (2018). This paper also includes a good discussion of scaffold level analysis, as well as predictions of increased drug success. A more mathematical general analysis of this is covered in Lovering et al. (2009). Monteleone et al. (2017) also concur with this, together with the presence of aliphatic rings and chirality, although antibacterials were not specifically analysed in their work.

The presence of nitrogen is also significant in clinical bioactives and an analysis of U.S. FDA approved drugs published in 2014 revealed that 59% of unique small molecule drugs contained a nitogen heterocycle (Vitaku et al. 2014). Such heterocycles included fused and bridged systems and medium ring or macrocyclic compounds, which we propose, further underscores the relevance of looking at nitrogen-containing tropane and berberine analogues in the context of new hybrid designs.

Tropane-derived analogues

The biologically active tropane alkaloids are an important natural product group and a number are available commercially as structurally defined starting materials (Kohnen-Johannsen and Kayser 2019). Some tropane alkaloids have also shown antibacterial activity (Özçelik et al. 2011). The tropane alkaloid scopolamine (Scheme 3.2a) also has some RND efflux pump blocking activity but displayed synergistic activity only with the antibacterial carbenicillin in *Escherichia coli* (Aparna et al. 2014). The tropane skeleton has also served as a useful starting point for the synthesis of an array of other systems and derivatives as illustrated by the work of Lowe et al. (2020), although not specifically for new antibacterials. It is how the tropane system might be used in various ways to access structures within the type (i) hybrid antibacterial manifold which is the focus here.

With the tropane derivatives, many opportunities present for the ready incorporation of pharmacophoric units in substituents attached to, or as part of, aza-bicyclic and tricyclic templates. Two of these readily available derivatives are tropenone

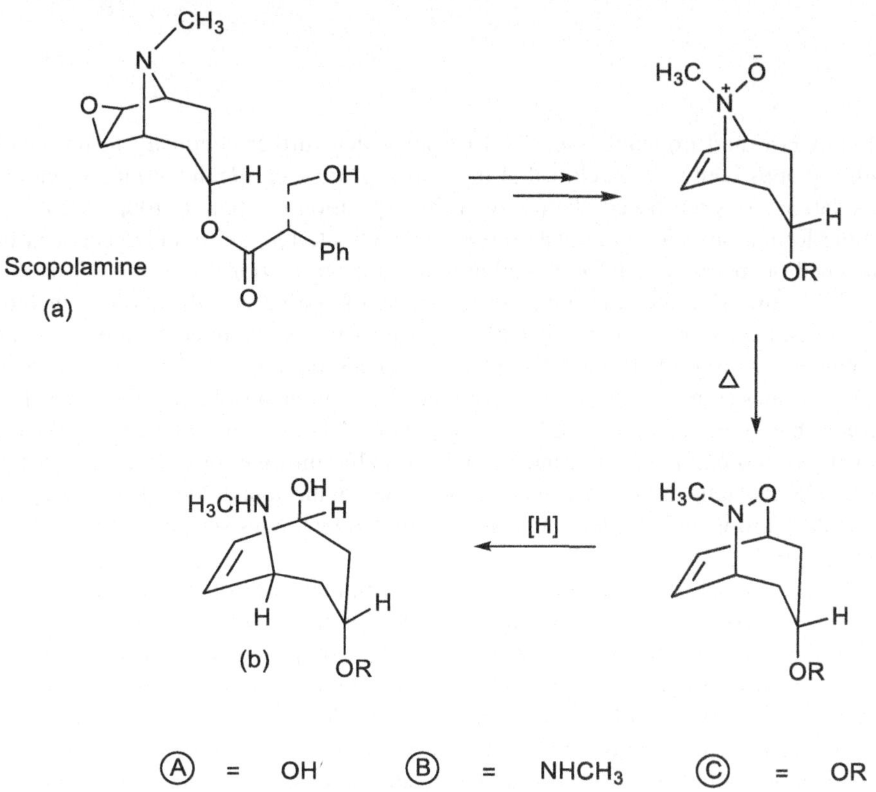

Scheme 3.2 Synthesis proposal for a potential hybrid (**b**) from scopolamine (**a**) with three possible target interaction sites and linking groups as part of the ring

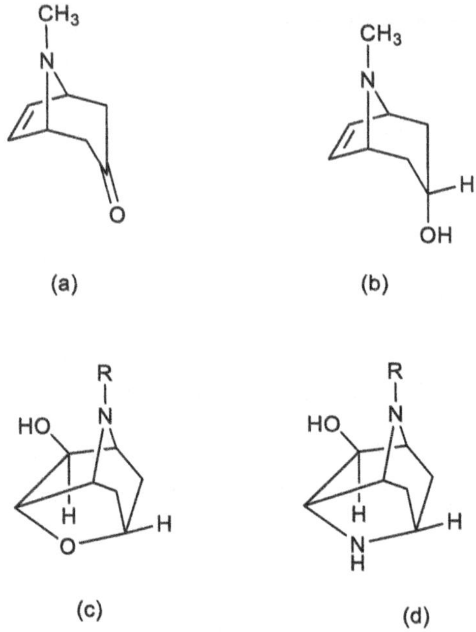

Fig. 3.21 Structures of some tropane derivatives: tropenone (**a**), tropenol (**b**) and other bridged systems (**c** and **d**)

(Fig. 3.21a) and tropenol (Fig. 3.21b) from which further chemistry to introduce other groups could be developed at the bridging nitrogen (dealkylation; acylation; alkylation), oxygen based chemistry at the 3-position (carbonyl group chemistry; imine formation; spiro ring elaboration to which site-recognition elements can be attached or incorporated; *O*-alkylation; enol ether formation) or at the 6,7-double bond. At this last site ring annelations are also possible and these could include substituted cyclopropanation via carbene chemistry or substituted cyclobutene ring annelation with gold (I) catalysis, which has been demonstrated for other unactivated alkenes (Bai et al. 2018). Further annelations might also be possible with electron rich enol ethers derived from the tropenone 3-keto group (or from the double bond reduced compound tropinone) and a suitable nitrogen protecting group. For example, aromatic or heteroaromatic ring formation through gold (I)-catalyzed [4 + 2] cycloaddition and elimination using substituted ynamides should be a possibility (Dateer et al. 2012).

Other established reactions can be used to manipulate the tropane skeleton via ring cleavage or via other ring construction reactions with the concomitant introduction of further, quite precisely positioned, functional groups. Illustrative of the former transformation one could use the [1,2] Meisenheimer rearrangement of *N*-oxides (Zhang 2011) followed by N–O bond reductive cleavage to access tri-substituted cycloheptenes (b) as shown in Scheme 3.2 with the potential pharmacophoric sites (or sites one could build on) noted. The synthesis of the specific compound (b, R = TBDMS, Scheme 3.2) from the tropane alkaloid scopolamine has been described earlier (Bremner et al. 1996). Enol ethers might also serve as precurors for the sequential Meisenheimer and reductive cleavage chemistry to afford other structural variations in the product. The Meisenheimer rearrangement should favour formation of a C1–O bond in this case and hence the positioning of the hydroxyl group after N–O bond cleavage.

Ring construction reactions from the tropane scaffold to give more rigid compact structures are also possible. For example, on basic hydrolysis of the ester side chain in scopolamine, intramolecular nucleophilic attack on the epoxide group occurs readily to give the tricyclic derivative oscine (Fig. 3.21c, R=CH_3). With this rigid skelton one has a tertiary hydroxyl substituent, and an ether oxygen and amino nitrogen as part of the ring system and pharmacophoric groups could be subsequently attached via this nitrogen or from the hydroxyl group. Additional possibilities for attachment of groups would be provided from the aza analogue (Fig. 3.21d, R=CH_3) of oscine, and derivatives, if this could be synthesised. A homologous system is known (DeCorte et al. 2018) but the 5-membered ring analogue does not appear to have been reported. Rigid polycyclic scaffolds in drug design, including multi-mechanism chimeras (Bansal and Silakari 2014; Van der Schyf and Geldenhuys 2009), are of continuing research interest.

Berberine and congeners

The biologically active, antibacterial quaternary alkaloid berberine (Figs. 3.9a and 3.12c) also provides much scope for the design of potential triple action agents of type i. It is available relatively cheaply commercially, and possesses a number of sites for structural tailoring and introduction of different groups. There are a number of reaction sites on the berberine skeleton and some of the key ones have been summarized in Sect. 3.2.3.2. Reactions at these sites provide good opportunities for the introduction of functional or structural moieties. For example substituent groups can be introduced at C13 in berberine, after conversion to dihydroberberine or 8-substituted dihydroberberines and subsequent re-conversion to the quaternary salt structure (Bremner and Kelso 2010). Also electrophilic substitution on berberine itself or on derivatives (for example berberrubine) enables the direct introduction of substituents at C12. With berberrubine this substitution occurred via a Mannich-type reaction (Li et al. 2014b; Mistry et al. 2017). Using these reactions, at each of these positions substituents incorporating different biological target motifs, A, or C (as in Fig. 3.22a or b), for example, could be introduced. The heterocyclic skeleton in the retained berberine core may act as a recognition element in itself (shown as B in Fig. 3.22a) for DNA binding (Jin et al. 2010) or inhibition of assembly of the protein FtsZ, a key protein for bacterial cell division (Boberek et al. 2010; Domadia et al. 2008). Domadia et al. (2008) include in their paper modelling studies on berberine interacting with FtsZ and from their results the inclusion of different small pharmacophores in the 9-OR group would seem possible, while maintaining FtsZ binding. Boberek et al. (2010) also indicate that this binding may be of greater importance than DNA binding in the antibacterial activity of berberine on the basis of genetic evidence in *Escherichia coli.* Various compounds show inhibitory activity against FtsZ including 3-methoxybenzamide (3-MBA) analogues (Stokes et al. 2013)

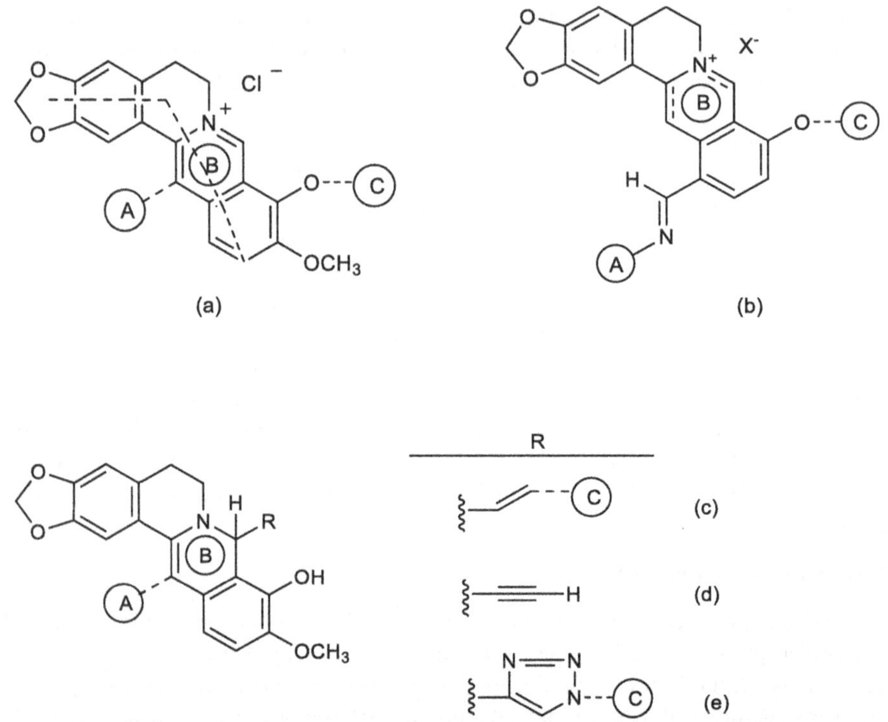

Fig. 3.22 Structures of some further hypothetical berberine and dihydroberberine derivatives (**a**–**e**) designed for multi-targeting

and benzofuroquinolinium derivatives, which are a new class of potent antibacterial agent (Zheng et al. 2018).

Berberine has moderate antibacterial activity by inhibiting the assembly of FtsZ, as do berberine analogues (Mori-Quiroz et al. 2018). In this paper the authors also indicate de novo synthesis of berberine analogues is underway and this represents an alternative approach to chemical modifications of berberine and related isoquinolinium alkaloids for SAR studies. Worth considering both approaches. The design of berberine-based FtsZ inhibitors which have broad spectrum antibacterial activity has also been reported by Sun et al. (2014). Two binding sites on FtsZ were found with one being the C-terminal interdomain cleft and the other the GTP-binding site. The potential for a number of FtsZ binding sites was also revealed in an *in silico* analysis by Kusuma et al. 2019 (Kusuma et al. 2019). Continuing this theme, an important recent review by Casiraghi et al. (2020) has summarized a range of FtsZ inhibitors classifying them according to their main protein binding sites, which is very useful from a medicinal chemistry perspective in enabling some future design ideas with respect to targeting a number of the protein sites.

Berberine can be demethylated in high yield to the corresponding 9-hydroxy analogue berberrubine, a compound which, although classified as an alkaloid, is usually an artifact of isolation from plant sources (Grycová et al. 2007). Other substituted berberines can also undergo this thermal demethylation process. The demethylation can be accomplished by straightforward thermolysis (Lo et al. 2013) providing a convenient hook for the attachment of further groups by reaction at the 9-OH site produced (Li et al. 2010). The 9-hydroxyl group can also possibly serve to direct further nucleophilic attack, for example by boronic acids in a Petasis type reaction (Guerrera and Ryder 2016) (Guerrera and Ryder 2016), to afford 8-substituted derivatives of the type shown in Fig. 3.22c and d. For systems of the latter type, standard azide-based click chemistry (Thirumurugan et al. 2013) could then be used to attach a unit with the recognition element C as in Fig. 3.22e.

While berberrubine itself displays only weak to very weak antibacterial activity against the Gram-positives *Micrococcus luteus* (Kim et al. 2002) and *Mycobacterium smegmatis* (Gharbo et al. 1973) greater potency through increased target interaction possibilities may result from conversion of the phenolic hydroxyl group at C9 in both compounds Fig. 3.22c and e to the fluorosulfate group using SuFEx click chemistry (Liu et al. 2018). Toxicity considerations in the handling of the initially required gaseous reagent for this reaction, sulfuryl fluoride, has been mitigated by the introduction of fluorosulfuryl imidazolium salts which are bench-stable and act as alternative donors of the '$F–O_2S^+$' unit. An excellent detailed review of SuFEx chemistry and applications in late-stage functionalization strategies for bioactive compounds has been published recently by Barrow et al. (2019). The introduced SO_2F group is also a useful handle to add other pharmacophores.

Another valuable reaction of berberine and congeners is the direct introduction of secondary amino functionality at the 9-position by an aromatic nucleophilic displacement reaction with primary amines. Although yields were moderate in the case of palmatine, the ease of the reaction is a positive attribute, and through this route, a derivative was found with good anti-*Helicobacter pylori* activity, with this possibly being mediated via inhibition of *Helicobacter pylori* urease (Fan et al. 2020). This displacement reaction has also been applied to berberine to give the 9-amino substituted derivatives (Naruto et al. 1976; Wang et al. 2018).

It is also relevant to note here that tetrahydroberberine derivatives obtained from berberrubine and with different substituents at the 9- and 12-positions have antibacterial activity against both Gram-positive and Gram-negative pathogens, particularly one derivative with a C9-benzyloxy group and an imino-1,3,4-triazolyl group at C12 (Fig. 3.23b). This derivative had in vitro MIC values of 2 $\mu g\ mL^{-1}$ and 8 $\mu g\ mL^{-1}$ against MRSA and *Pseudomonas aeruginosa* respectively (Duan et al. 2017). Dual DNA-targeting was indicated. with the 1,3,4-triazole moiety present being involved (on the basis of molecular modeling studies) but this ring might also facilitate interaction with other targets. The key intermediate in the synthesis of this triazole derivative and congeners was the tetrahydroberberrubine-12-aldehyde (Fig. 3.23a), which was accessible in turn from berberine chloride via a compact three step synthesis in good overall yield. This aldehyde, plus the 9-phenolic hydroxyl group, provide good opportunities for the elaboration of further pharmacophoric elements at these sites with a view to designed multiply active antibacterials.

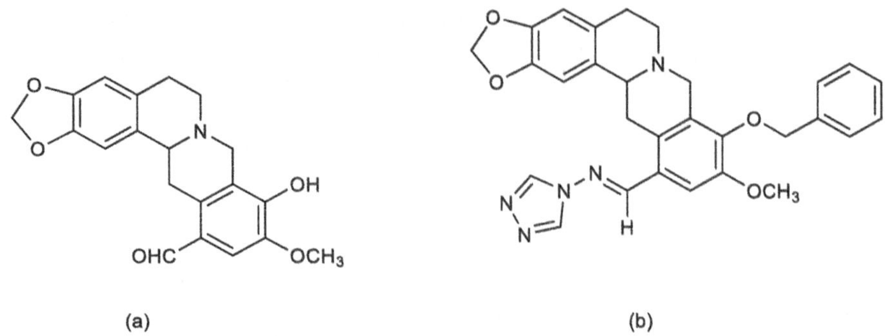

Fig. 3.23 Structure of tetrahydroberberrubine-12-aldehyde (**a**) and the antibacterial triazolo derivative (**b**)

The aldehyde (Fig. 3.23a) was also a key precursor in the synthesis of a related derivative with a C=C linkage to a nitroimidazole unit (Zhang et al. 2018). Berberine-based nitroimidazole derivatives, for example the potent derivative shown below (Fig. 3.24), may be multi-targeting against drug-resistant *Escherichia coli* from an analysis of possible binding sites (Zhang et al. 2018). The substituents aided multi-targeting involving possible binding to DNA polymerase III (affecting DNA synthesis), DNA intercalation (affecting transcription), and membrane permeabilization. Consistent with multi-targeting, resistance development in *Escherichia coli* was strongly disfavoured on the basis of the in vitro multi-passaging results. The compound shown in Fig. 3.24 was more active than norfloxacin or berberine in vitro.

A facile synthetic entry to the tetrahydroprotoberberine and protoberberine systems involving either oxalyl chloride-induced cyclization of protopine (or allocryptopine) or acid-catalyzed cyclization of the dihydro derivative of protopine has been described together with data on the generally modest in vitro antibacterial activity of these compounds, although berberine was somewhat more active against *Staphylococci* (Cheng et al. 2014).

In addition to the above sites for the attachment of substituents, the fused methylenedioxy group in berberine can be selectively ring opened giving access to a 2,3-diol (demethyleneberberine) on reaction with boron tribromide in dry dichloromethane at 0 °C for 1 h; if this mixture was then heated at reflux for a

Fig. 3.24 Structure of a multi-targeting nitroimidazole-tetrahydroberberrubine derivative

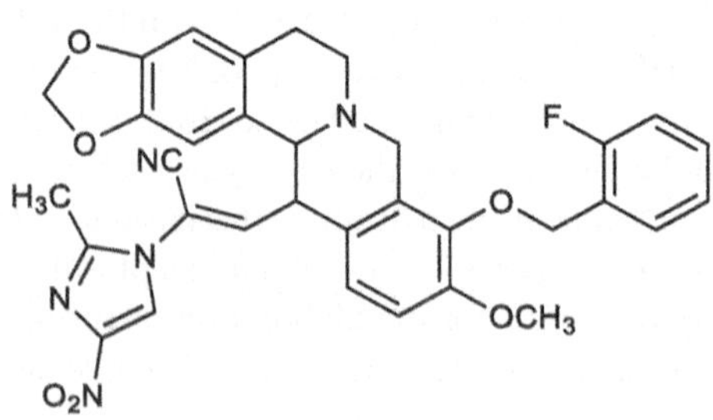

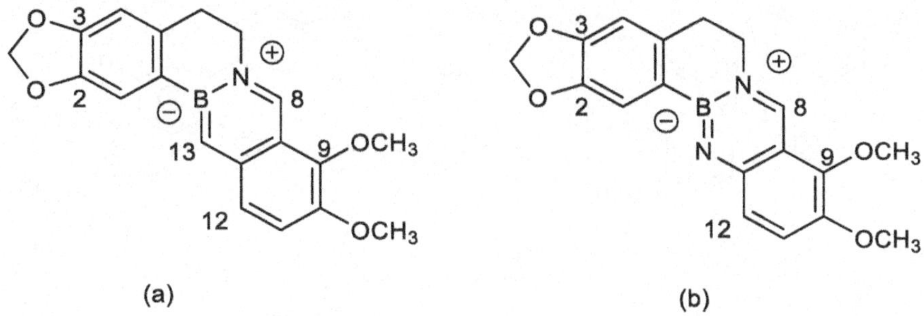

Fig. 3.25 Structures of a proposed aza-bora system (**a**) and diaza-bora system (**b**) based on berberine

further 20 h the two methoxy groups present in berberine could also be demethylated (Roselli et al. 2016). The catecholic 2,3-diol from the low temperature reaction could serve in turn as anchor points for other groups, although regioselectivity issues may be a problem if different groups are involved. Furthermore the catechol moiety could act as a siderophore for the Trojan horse uptake approach into the bacterial cell.

To change the electronic properties of berberine one option is to change the skeletal atoms, as in the hypothetical boron analogue shown in (Fig. 3.25a). With this borazine-type structure the N and B sites are isolobal respectively to cationic and anionic CH fragments making them interesting units for antibacterial design architectures (Islas et al. 2007). Overall the compound is neutral but with charged atomic loci. This could perhaps result in evasion of efflux pumps (like the NorA pump to which the charged berberine unit is susceptible) while still retaining good penetration and bacterial DNA/FtZ binding properties as well as minimizing molecular weight and size increases.

An extension of the design to replace the 13-CH with N in the BN compound (Fig. 3.25a) to give the diazabora system (Fig. 3.25b) might also be worth considering in this context. While no synthetic schemes are proposed at this point, it is of interest to note that a BN-pyrene derivative with a central $N^{+}{=}B^{-}$ double bond is readily made and is stable (Jaye et al. 2017). The linked attachment of other pharmacophoric groups to the basic core structures in the boron containing systems would also need to be considered in any synthetic proposals.

Further work would be justified on berberine-like lactam analogues with an 8-oxo group with the zwitterionic resonance contributing structure likely to be significant and their overall neutrality may enable better Gram-negative outer membrane penetration but with similar likely active site binding characteristics (Fig. 3.26a). Other pharmacophoric elements could be incorporated in A and C. In addition, by introducing B-OR in place of the lactam carbonyl (Fig. 3.26b) there is potential for incorporation of another pharmacophoric group on the B–O oxygen while having a formal negative charge on boron and a formally positively charged nitrogen. Additionally the berberrubine analogues would also provide extra linkage sites for such groups if required.

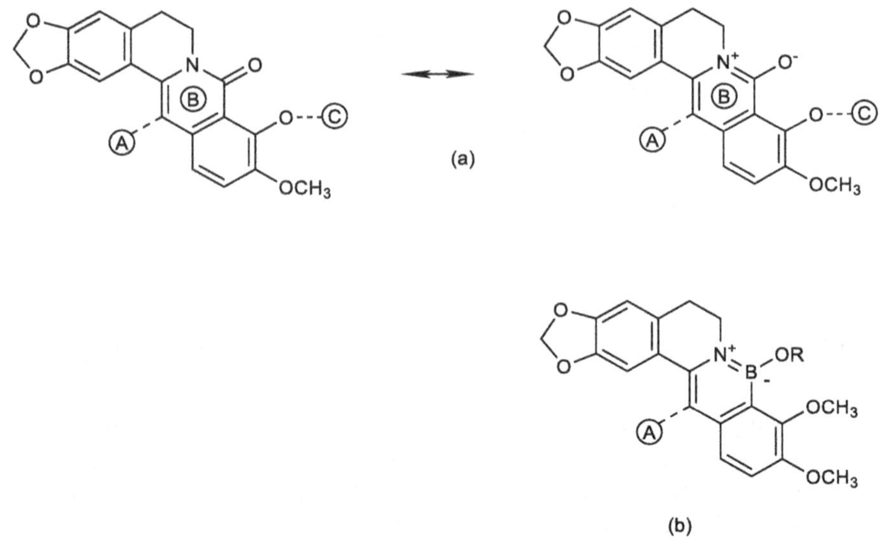

Fig. 3.26 Hybrid structure proposals from 8-oxodihydroberberine (**a**) and a related boron-substituted analogue (**b**)

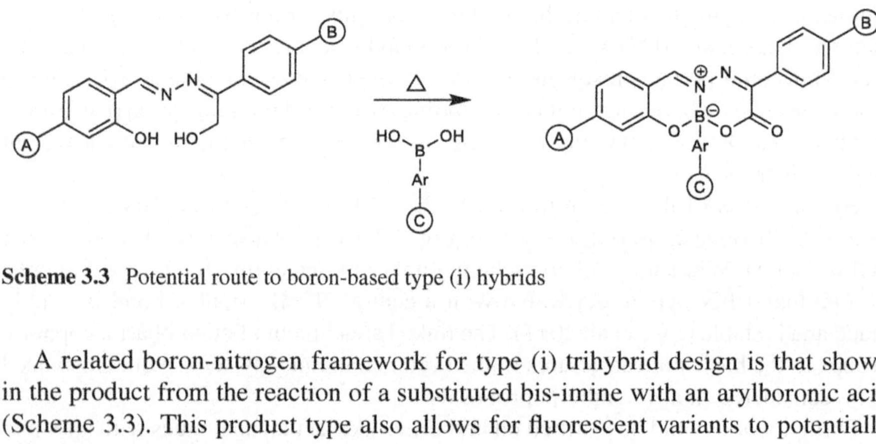

Scheme 3.3 Potential route to boron-based type (i) hybrids

A related boron-nitrogen framework for type (i) trihybrid design is that shown in the product from the reaction of a substituted bis-imine with an arylboronic acid (Scheme 3.3). This product type also allows for fluorescent variants to potentially afford access to more detailed target tracking. Such compounds are based on those described by Alcaide et al. (2017) and the synthesis should accommodate a range of pharmacophoric groups for different bacterial target sites.

3.3.4.3 Type II. A---BC

This type is distinguished by only one multi-atom linking group being present between A and BC, with the other two recognition units B and C being connected via systems with a common atom in a ring, or at least two common atoms in fused rings,

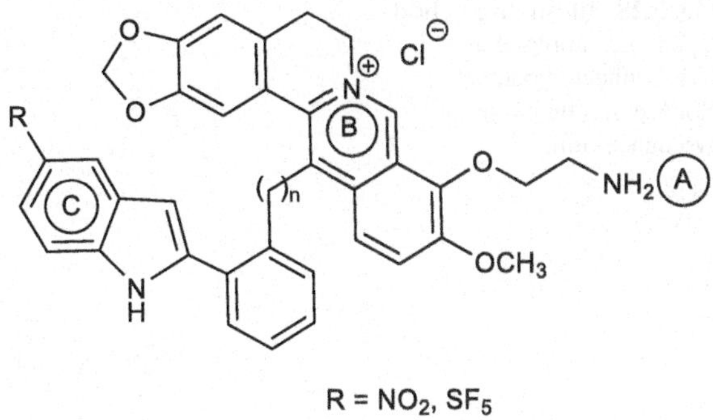

Fig. 3.27 Further molecular proposals for a berberine-derived type (i) template with three targeting sites

or one bond in a direct connection. Again, using berberine as the starting design point, one might envisage for example, attaching a NorA efflux pump blocking moiety, the 5-nitro-2-phenyl-1*H*-indole unit (C), directly (Fig. 3.27, n = 0) to the berberine core B to define the linked hybrid BC and this composite unit then may also be attached to a primary amino group as the third recognition element A at the 9-position via some suitable linker. If a 1-atom methylene group is used to join units C and B (n = 1 in Fig. 3.27) then formally one would classify the system in type (i) as two linking groups would then be involved overall. The primary amino group in A may also assist with the penetration of Gram-negative bacteria.

Further exploration of other functional groups as bioisosteres (Meanwell 2011) in the efflux pump blocker moiety, and other such moieties, could also be fruitful. In particular the pentafluorosulfanyl group (SF_5) as a bioisostere for the nitro group is pertinent. This group is chemically stable, and a strong electron withdrawing group, but unlike the nitro group, is quite lipophilic. The SF_5 group can also alter pKa and metabolic stability, although sometimes the SF_5 exchange for NO_2 is not always beneficial in terms of biological activity possibly due to the increased steric demands of the former group (Sowaileh et al. 2017). However the novel combination of properties evident with the SF_5 group has meant it is being explored more extensively in drug design. Safety and other issues in the synthetic access to this group are also being addressed, for example in a versatile gas-reagent free approach (Pitts et al. 2019).

Other ways to incorporate the NorA pump inhibitor unit directly linked to the berberine core might for example extend to the C12 position with a structure like that in Fig. 3.28 with the A unit attached by a linker at C13. Analogues of this structural type with ciprofloxacin (as unit A) attached as a 13-OR derivative with a N–C–O linkage to the piperazine NH group in ciprofloxacin should also be readily accessible. Similarly the ciprofloxacin piperazine moiety could be linked via a 13-C-N unit using the Cp*IR-catalysed N-alkylation of amines with alcohols described by Fujita et al. (2008), or via a more direct N–O linkage, although the latter may be susceptible to reductive cleavage by reductases.

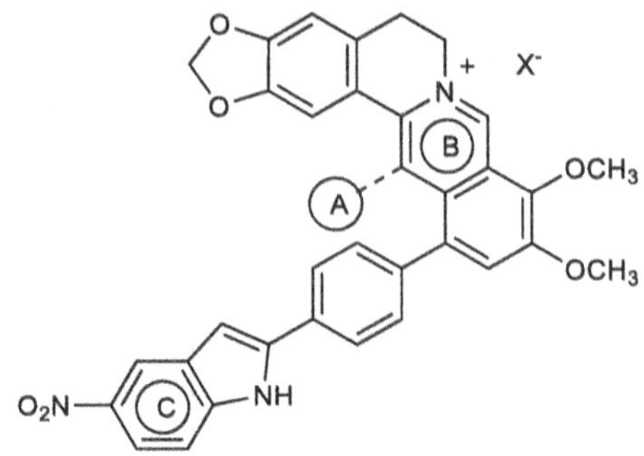

Fig. 3.28 Illustrative hybrid type (ii) example of a C12-connected potential NorA pump blocking (or avoidance) unit

3.3.4.4 Type III. A---B/C

Design type iii is a powerful one which represents an extension of type ii where B and C are partly or almost completely merged, or overlap spatially, as indicated for simplicity by the '/' symbol. In B/C if all the atoms nearly overlap then B and C would be close to equivalent but they are still designated as B and C since they could interact separately in different binding modes to two sites on one biological target or to two different biological targets, either mode being different from the target of A. Hence the triple action classification would still apply. This design is manifested in molecular terms in the antibacterial compound PD-2b (Fig. 3.29a) (Jayaraman et al. 2013). This compound incorporates a linked combination of a protocatechuic acid unit at one end and a sulfadiazine unit at the other. The hybrid compound was shown to be capable of targeting both DNA gyrase subunit B and topoisomerase IV subunit B (of *Pseudomonas aeruginosa*) from computer–based docking studies, as well as dihydrofolate reductase and is a promising potential antibacterial lead structure for further development. DNA gyrase is a tetrameric enzyme with two GyrA subunits and two GyrB subunits and is a sub-class of Type II topoisomerase, while topoisomerase IV is the other Type II topoisomerase in bacteria. While gyrase and topoisomerase IV are related and have similarities with the amino acid sequences they are different and have different roles. Gyrase is not present in higher eukaryotes making it a good target for antibacterial agents.

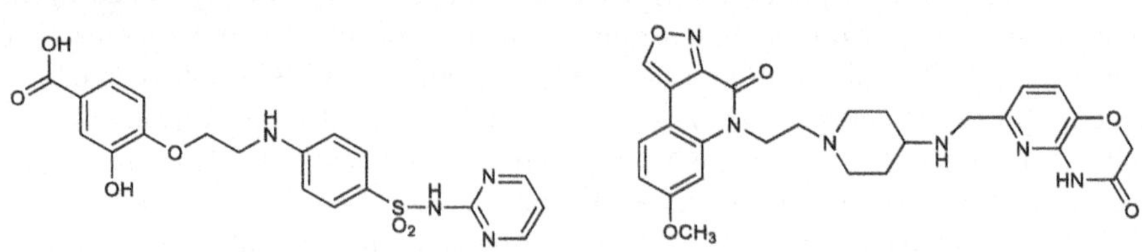

Fig. 3.29 Structures of the DNA gyrase and topoisomerase (IV) inhibitors PD-2b (**a**) and REDX 06213 (**b**)

The structurally different bacterial topoisomerase inhibitor REDX 06213 (Fig. 3.29b) can also be considered under this general structural type iii (Charrier et al. 2017). This compound, along with others in this group, features an isoxazoloquinolone group at one end which interacts with DNA (A), a piperidine-based linking group, and then a pyrido-oxazinone structural motif (B/C) at the other end which binds to bacterial gyrase and to topoisomerase IV enzymes. REDX 06213 showed good antibacterial activity with MICs in the low micromolar range against multi-drug resistant strains of the Gram-negatives *Escherichia coli* and *Acinetobacter baumannii* and with a low likelihood for the development of resistance.

Considerable scope exists for further extension of the type iii triple action agent design to berberine-based analogues which might include structures like those in Fig. 3.30 (a–d). For example systems like those shown in Fig. 3.30a and b, with the heterocyclic core represented by B potentially binding to bacterial DNA and the recognition unit incorporated in the linked group A interacting separately with a gyrase or topoisomerase IV enzyme. The fused 5-nitroindolic unit C could act as a recognition unit to interfere with efflux of the whole quaternary salt by, for example, the NorA efflux pump. The proposed four-ring fused heteroaromatic skeleton in

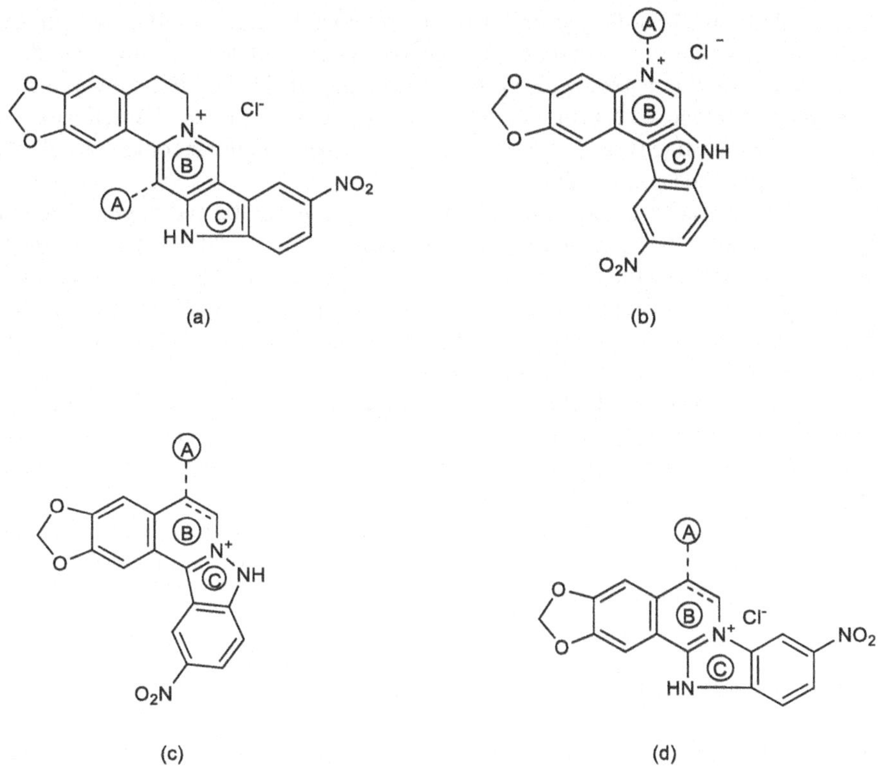

Fig. 3.30 Structures of four suggested berberine-based chimeric-type hybrids (**a**–**d**)

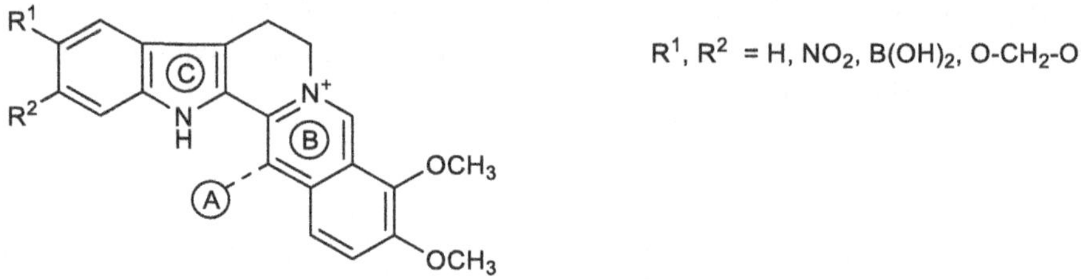

Fig. 3.31 Benzene ring replacement by an indolic unit to form a type (iii) A---B/C hybrid

Fig. 3.30b (a 7*H*-indolo[2,3-*c*]quinoline, excluding the methylenedioxy group) is isomeric with those present in the bioactive indoloquinoline alkaloids (Bogányi and Kámán 2013; Etukala et al. 2008); antibacterial activity in these alkaloids has been described together with the activities of analogues (Lavrado et al. 2010). However, cytotoxicity may be a problem with these systems as a result of DNA intercalation (Van Miert et al. 2005).

The heterocyclic core structure in Fig. 3.30a is isomeric with the readily accessible tetrahydro-6*H*-benzo[*a*]indolo[2,3-*g*]quinolizines (Guo et al. 2016) and the same synthetic methodology used for these compounds could be adapted to make the heterocyclic core in the proposed isomeric system followed by ring oxidation to give the berberine analogue derivatives. Similarly, fusion reactions involving *o*-aminophenylboronic acids (Tumey et al. 2014) to access the isomeric skeleton to that in the indoloquinoline (Fig. 3.30b) have been described and these could possibly be adapted to give the latter system.

Variations on the structural themes expressed in Figs. 3.30a and b with different N-positions and/or ring size and ring fusion arrangements could also be of interest. There are many possibilities but one is shown in Fig. 3.30d which incorporates the known 12*H*-benzo[4,5]imidazo[2,1-a]isoquinolin-7-ium system (Zhu et al. 2017). The alternative ring fusion would afford the indazolo system represented by Fig. 3.30c and with elements of a pump blocking motif.

Other variations on the berberine pattern might include the more classical incorporation of fused ring isosteres like thiophene or pyrrole units for a benzene ring with a methylenedioxy ring fused to the isosteric unit. Alternatively, other heteroaromatic ring replacements like an indole unit (C), could also incorporate in part some similarity to the NorA efflux pump inhibitor INF55 or to the berberine DNA recognition template (B) (Fig. 3.31). The linked unit A could then provide possible binding opportunities with another target site in this design.

3.3.4.5 Type IV. ABC

Typical of this design type, and of the following types v and vi, is the absence of any extra linking groups. Thus for type iv, the three recognition elements could each be joined directly via one or two common atoms (each in spiro centres) and one or two common bonds including those embedded in ring systems as in fused systems. The recognition elements could be incorporated in a ring and/or be included in substituent groups. There does not appear to be any clear precedent in the literature of this general type of triple action antibacterial agent, but possibilities for the future based on the berberine template might include for example recognition features A and C being directly attached to berberine (recognition unit B) perhaps at two different positions such as at the 8 and 12, 8 and 13, or 12 and 13 berberine positions.

3.3.4.6 Type V. AB/C

In the representation of this type, B/C would indicate an overlap connection between the two as for BC in type iii. The key differentiating characteristic of this type v is the absence of extra linking atom groups for A toB/C, but these linking groups would be replaced by connection via a common atom in a ring (or not), or at least two common atoms in fused rings, or one bond in a direct connection. B/C represents partial up to near complete spatial and structural overlaps of the other two recognition elements B and C. For the near complete overlap case B would then be close to equivalent to C and action at two biological targets could be envisaged as noted for type iii. Prior exploration of this type in the berberine context does not seem to have been reported, but if A was directely attached to B/C in structures like those in Fig. 3.30a and b, then they would represent such an expression of the type v design concept. Such a structure based on the former would still have planar ring fusions to give a similar bent geometry and a positively charged quaternary nitrogen as in berberine.

While not entirely clear, one can possibly include the promising antibacterial diazabicyclooctane derivatives OP0595 (Fig. 3.32a) (Morinaka et al. 2015; Livermore et al. 2016), and zidebactam (Fig. 3.32b) (Livermore et al. 2017) in the type v group. The core diazabicyclooctanone unit in these compounds behaves like a β-lactam unit; class A and class C β-lactamases are inhibited by OP0595 which also has good antibacterial activity against a number of Gram-negative pathogenic bacteria through strong binding to the penicillin-binding protein PBP2. In addition, it acts as an "enhancer" of the antibacterial activity of β–lactams which bind to other

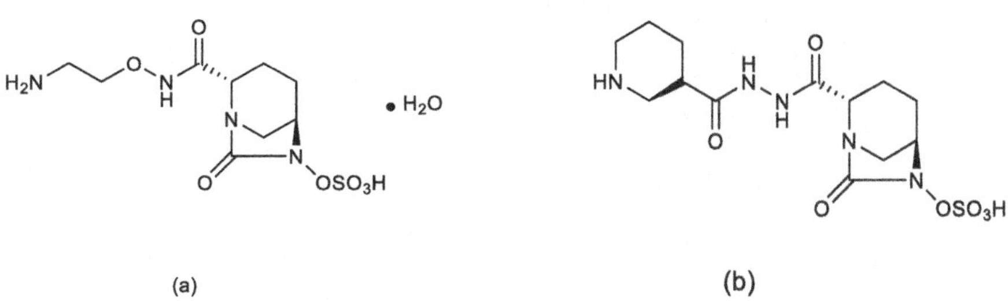

Fig. 3.32 Structures of OP0595 (**a**) and zidebactam (**b**)

penicillin-binding proteins, independent of the blocking of the β-lactamases (Livermore et al. 2016; Morinaka et al. 2016). Zidebactam has very good activity against some Gram-negative bacterial strains (Livermore et al. 2017; Sader et al. 2017). Analogue development to include new N–$B(OH)_2$ or N–O–$B(OH)_2$ units relacing the N–OSO_3H functionality could also be worth further consideration if resistance starts to appear with OP0595 and zidebactam. However stability to hydrolysis may be an issue with these functional groups and this would need to be carefully checked under physiological conditions.

3.3.4.7 Type VI. A/B/C

Taking the type v design further one would then arrive at the condensed structural type vi, in which a partial up to complete or near complete spatial overlap of the recognition elements could be envisaged. With this type it is considered less likely to be complete overlap of all three units in one molecule interacting with similar binding strengths at three different biological targets with the same recognition element.

While not apparently reported as yet, one can also envision compounds of type vi with partial structural overlaps and with a prospective triple action profile. Illustrative of this molecular design, part of another proposed structure (Fig. 3.30b) could be merged with a flavanone unit A to give the structure as shown (Fig. 3.33a). The merged unit A could potentially mimic the flavonoid moiety present in the dimeric flavanone, tetrahydroamentoflavone (Fig. 3.33b) (Muhs et al. 2017). Compounds of this type have been noted as present in an extract from the Brazilian peppertree, *Schinus terebinthifolius* Raddi (Anacardiaceae). The flavone-rich extract from this tree showed inhibitory activity against the accessory gene regulator (agr) alleles in *Staphylococcus aureus* without growth inhibition (Muhs et al. 2017), thus suppressing quorum sensor expression and quorum sensing controlled virulence factors. The plant extract showed very good activity in vivo in the treatment of dermonecrosis caused by a virulent strain of MRSA. While the biological assessment of pure components from the

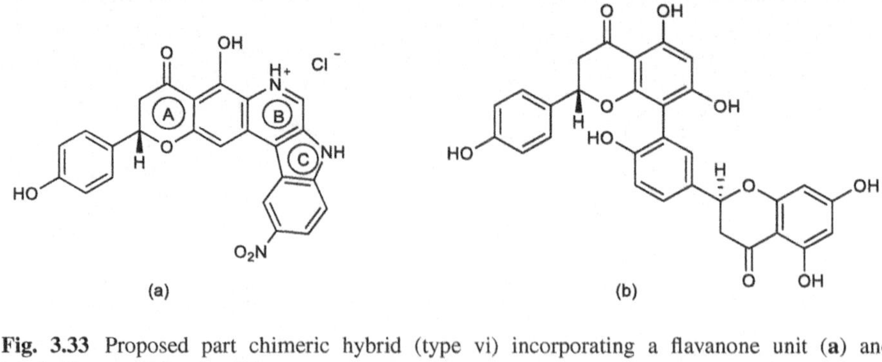

Fig. 3.33 Proposed part chimeric hybrid (type vi) incorporating a flavanone unit (**a**) and tetrahydroamentoflavone (**b**)

extract needs to done, tetrahydroamentoflavone is one of the significant components in the active extract, and it is thus conceivable that incorporating elements of this flavonoid structure in the hypothetical structure (Fig. 3.33a) might also show quorum quenching activity. With such a compound there could thus be a DNA focus for two of the interactions (A/B portion) while C might interfere with the NorA pump-mediated efflux of the molecule in *Staphylococcus aureus*. Indicative assessment of the NorA pump inhibitory activity would have merit in this process in order to highlight the most likely designs to be of interest. Such assessments might usefully involve in silico docking studies on a homology model of the pump for which an X-ray structure is not currently available (Gupta et al. 2018; Zimmermann et al. 2019).

On a general note, other groupings of target recognition elements can be envisaged which are not covered by types i–vi but which would extend the possible design possibilities for future studies. By way of illustration, one could for example consider structural types involving just two recognition elements, A and B, but with one having two different binding sites and the second a different site for the triple activity. The units A and B might then be linked by one or more extra linking atoms (A---B) or be directly linked or fused or be linked by a common atom (AB). Further progression of the fusion would then lead to condensed systems represented by A/B.

3.4 More Than Triple Action Hybrid Agents

One would be justified in asking whether incorporation of more than three separate recognition sites in the one hybrid molecule was a pipedream or was it a likely possibility for multi-modal single molecule antibacterials? This author considers that it is likely to be the latter on the basis of such intentionally designed agents with potentially more than three separate activities in other areas being described, for example, in the development of new antipsychotic drugs (Zajdel et al. 2018). Hybrids were deliberately designed for interactions with greater than three receptor or receptor sub-type sites and included structural elements known to favour selective interactions with monoaminergic receptors; both agonistic and antagonistic interactions were observed depending on the receptor, together with SERT blockade in a selected instance.

But there are inherent limitations in the design of such hybrids. Achieving an efficacious concentration of the antibacterial at each target site is more problematic, although it can be achieved as evidenced by the re-purposed drug auranofin (see Chap. 1, Fig. 1.4 for the structure), which shows promising multi-site activity against Gram-positive pathogens (Thangamani et al. 2016), and against *Mycobacterium tuberculosis* in vitro (Harbut et al. 2015). Auranofin displays a complex interplay of activities involving targets in the cell wall, DNA and bacterial protein synthesis, but outer membrane permeability issues, together possibly with efflux pump activity (e.g., AcrAB activity), compromised Gram-negative efficacy and further structural adjustments may be necessary to increase penetration; alternatively, drawing on other work, pentamidine may be worth adding as a membrane disruptor (Stokes et al.

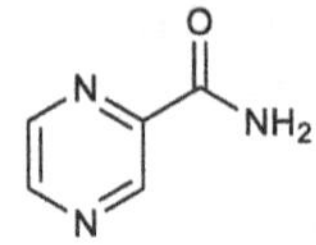

Fig. 3.34 Pyrazinamide

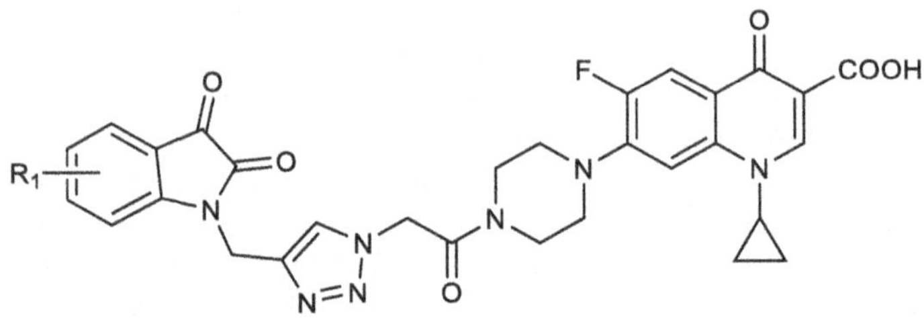

Fig. 3.35 Structures of two ciprofloxacin-triazole-isatin hybrids

2017) as long as this does not also disrupt the gold complexation in a negative way. Interestingly, the inclusion in the culture broth at a low (sub-inhibitory) concentration of polymixin B-nonapeptide, a permeabilizing agent, dramatically increased the potency of auranofin against a wide range of Gram-negative bacteria (Thangamani et al. 2016). The small molecule therapeutic pyrazinamide (Fig. 3.34) also inhibits multiple targets in *Mycobacterium tuberculosis* (Zhang et al. 2014; Lirin and Aparna 2016).

Ciprofloxacin-1,2,3-triazole-isatin hybrids (Fig. 3.35, R_1 = H, 5-F) have, potentially, at least four target binding sites with two enzyme targets for the ciprofloxacin moiety and with the isatin group being susceptible to separate nucleophilic attack by probably more than one nucleophilic group on other bacterial proteins. Very good in vitro activity was seen with one of these compounds against both drug-susceptible, and multidrug resistant, *Mycobacterium tuberculosis*, but unfortunately cytoxicity against Vero cells was also observed, possibly due, at least in part, to the potentially promiscuous activity of the isatin component in this hybrid. Further structural modification investigations with these hybrids could be rewarding (Chen et al. 2019).

There is considerable further scope for multi-targeted hybrid design using the piperazine NH in ciprofloxacin as a point of attachment and keeping the key carboxylic acid group unencumbered. For example, one might also include a salicylidene acylhydrazide pharmacophore which alone (Fig. 3.36), or as its gallium (III) complex, is likely to interact with over three different targets (Hakobyan et al. 2014). The hydrazido carbonyl group and the phenolic hydroxyl in this derivative are well positioned to provide good binding to Ga (III), or to Fe(III) in situ.

Other multi-targeted approaches have been used successfully by Bortolotti et al. (2019). In this work, LasR quorum-sensing inhibitors were conjugated with

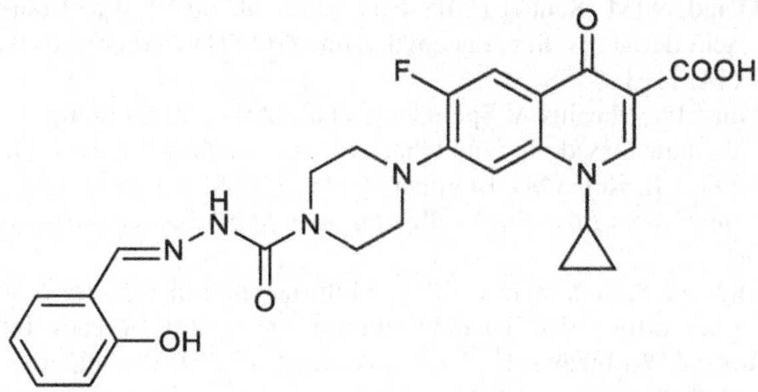

Fig. 3.36 Proposed ciprofloxacin-acylhydrazide hybrid design for multi-targeting

ciprofloxacin via an amide linkage at the piperazine NH to decrease the antibiotic tolerance of *Pseudomonas aeruginosa.* Such hybrids presumably are acting through at least three target sites.

Another possibility in the multi-target context is the development of semisynthetic antimicrobial peptides (AMPs). Natural AMPs seem to operate through targeting multiple hydrophobic sites and polyanionic sites or just these latter sites (Fjell et al. 2012; Torres et al. 2018, and references therein). The range of antimicrobial actions of AMPs provide a good guide for the development of more efficacious multiply active analogues with a low propensity to induce resistance. Computer-assisted design paradigms show promise in this area and are likely to be more widely applied in developing new peptidic systems (Fjell et al. 2012).

Synthetic cyclic liponapeptides also look very interesting for the control of Gram-negative pathogens. Multi-target interaction sites have been suggested for these polymyxin analogues, which also seem to have the potential to disengage the problematic nephrotoxicity seen from the antibacterial activity (Gallardo-Godoy et al. 2019). The related octapeptin antibiotics appear to differ in their mode of action and are promising 'rediscovered' leads (Blaskovich et al. 2018b; Blaskovich et al. 2019).

Notwithstanding all these promising actual and potential developments with multi-action/targeting hybids, it will probably be more fruitful to look at possible prodrug approaches to attack more than three sites and these are discussed, along with other considerations, in the following Chap. 4.

References

Abdali N, Parks JM, Haynes KM et al. (2017) Reviving antibiotics: efflux pump inhibitors that interact with AcrA, a membrane fusion protein of the AcrAB-TolC multidrug efflux pump. ACS Infect Dis 3 (1):89–98

Abrusán G, Marsh JA (2019) Ligands and receptors with broad binding capabilities have common structural characteristics: an antibiotic design perspective. J Med Chem 62 (21):9357–9374

Al Shaer D, Al Musaimi O, de la Torre BG et al. (2020) Hydroxamate siderophores: Natural occurrence, chemical synthesis, iron binding affinity and use as Trojan horses against pathogens. Eur J Med Chem 208:112791

Alcaide MM, Santos FMF, Pais VF et al. (2017) Electronic and functional scope of boronic acid derived salicylidenehydrazone (BASHY) complexes as fluorescent dyes. J Org Chem 82 (14):7151–7158

Amaral L, Martins A, Spengler G et al. (2014) Efflux pumps of Gram-negative bacteria: what they do, how they do it, with what and how to deal with them. Front Pharmacol 4:168–168

Ambrus JI, Kelso MJ, Bremner JB et al. (2008) Structure-activity relationships of 2-aryl-1*H*-indole inhibitors of the NorA efflux pump in *Staphylococcus aureus*. Bioorg Med Chem Lett 18:4294–4297

Anusuya S, Natarajan J (2012) Multi-targeted therapy for leprosy: in silico strategy to overcome multi drug resistance and to improve therapeutic efficacy. Infect Genet Evol 12 (8):1899–1910

Aoki T, Yoshizawa H, Yamawaki K et al. (2018) Cefiderocol (S-649266), A new siderophore cephalosporin exhibiting potent activities against *Pseudomonas aeruginosa* and other gram-negative pathogens including multi-drug resistant bacteria: Structure activity relationship. Eur J Med Chem 155:847–868

Aparna V, Dineshkumar K, Mohanalakshmi N et al. (2014) Identification of natural compound inhibitors for multidrug efflux pumps of *Escherichia coli* and *Pseudomonas aeruginosa* using in silico high-throughput virtual screening and in vitro validation. PloS One 9 (7):e101840

Aswathanarayan JB, Vittal RR (2018) Inhibition of biofilm formation and quorum sensing mediated phenotypes by berberine in *Pseudomonas aeruginosa* and *Salmonella typhimurium*. RSC Adv 8:36133–36141

Baell JB, Holloway GA (2010) New substructure filters for removal of pan assay interference compounds (PAINS) from screening libraries and for their exclusion in bioassays. J Med Chem 53:2719–2740

Baell JB (2016) Feeling nature's PAINS: Natural products, natural product drugs, and pan asay interference compounds (PAINS). J Nat Prod 79(3):616–628

Baell JB, Nissink JWM (2018) Seven year itch: Pan-assay interference compounds (PAINS) in 2017-utility and limitations. ACS Chem Biol 13:36–44

Bai Y-B, Luo Z, Wang Y et al. (2018) Au-catalyzed intermolecular [2+2] cycloadditions between chloroalkynes and unactivated alkenes. J Am Chem Soc 140:5860–5865

Ball AR, Casadei G, Samosorn S et al. (2006) Conjugating Berberine to a Multidrug Resistance Pump Inhibitor Creates an Effective Antimicrobial. ACS Chemical Biology 1 (9):594–600

Barrow AS, Smedley CJ, Zheng Q et al. (2019) The growing applications of SuFEx click chemistry. Chem Soc Rev 48 (17):4731–4758

Batra N, Rajendran V, Agarwal D et al. (2018) Synthesis and Antimalarial Evaluation of [1, 2,3]-Triazole-Tethered Sulfonamide-Berberine Hybrids. ChemistrySelect 3:9790–9793

Ben W, Qiaoyun Z, Wenzhu Y et al. (2018) In-vitro characterisation of a novel antimicrobial agent, TNP-2092, against *Helicobacter pylori* clinical isolates. Swiss Med Wkly 148:w114630

Beno BR, Yeung K-S, Bartberger MD et al. (2015) A survey of the role of noncovalent sulfur interactions in drug design. J Med Chem 58 (11):4383–4438

Beppler C, Tekin E, Mao Z et al. (2016) Uncovering emergent interactions in three-way combinations of stressors. J R Soc Interface 13 (125)

Besnard J, Ruda GF, Setola V et al. (2012) Automated design of ligands to polypharmacological profiles. Nature 492 (7428):215–220

Bhowmik D, Buzzetti F, Fiorillo G et al. (2014) Synthesis of new 13-diphenylalkyl analogues of berberine and elucidation of their base pair specificity and energetics of DNA binding. MedChemComm 5 (2):226–231

Blanco P, Hernando-Amado S, Reales-Calderon JA et al. (2016) Bacterial multidrug efflux pumps: much more than antibiotic resistance determinants. Microorganisms 4 (1):14

Blanco P, Sanz-García F, Hernando-Amado S et al. (2018) The development of efflux pump inhibitors to treat Gram-negative infections. Expert Opin Drug Discov 13 (10):919–931

Blaskovich MAT, Hansford KA, Gong Y et al. (2018a) Protein-inspired antibiotics active against vancomycin- and daptomycin-resistant bacteria. Nat Commun 9 (1):22

Blaskovich MAT, Pitt ME, Elliott AG et al. (2018b) Can octapeptin antibiotics combat extensively drug-resistant (XDR) bacteria? Expert Rev Anti Infect Ther 16 (6):485–499

Boberek, J. M., Stach, J., and Good, L. 2010. Genetic evidence for inhibition of bacterial division protein FtsZ by berberine. PloS One 5(10): e13745

Bogányi B, Kámán J (2013) A concise synthesis of indoloquinoline skeletons applying two consecutive Pd-catalyzed reactions. Tetrahedron 69 (45):9512–9519

Bolognesi M L (2013) Polypharmacology in a single drug: Multitarget drugs. Curr Med Chem 20: 1639–1645.

Bolognesi M L (2019) Harnessing polypharmacology with medicinal chemistry. ACS Med Chem Lett 10:273–275

Borges A, Saavedra MJ, Simões M (2015) Insights on antimicrobial resistance, biofilms and the use of phytochemicals as new antimicrobial agents. Curr Med Chem 22 (21):2590–2614

Bradford PA, Miller AA, O'Donnell et al. (2020) Zoliflodacin: An oral spiropyrimidinetrione antibiotic for the treatment of *Neisseria gonorrheae*, including multi-drug-resistant isolates. ACS Infect Dis 6:1332–1345

Breda S, Reva ID, Lapinski L et al. (2006) Infrared spectra of pyrazine, pyrimidine and pyridazine in solid argon. J Mol Struct 786 (2):193–206

Bremner JB, Browne EJ, Gunawardana IWK (1982) Oxaza and dioxaza medium-sized heterocycles and benz-fused derivatives. Heterocycles 19 (4):709–742

Bremner JB, Smith RJ, Tarrant GJ (1996) A Meisenheimer rearrangement approach to bridgehead hydroxylated tropane alkaloid derivatives. Tetrahedron Lett 37(1):97–100

Bremner JB, Ambrus JI, Samosorn S (2007) Dual-action based approaches to antibacterial agents. Curr Med Chem 14(13):1459–1477

Bremner JB, Keller PA, Pyne SG et al. (2010a) Binaphthyl-based dicationic peptoids with therapeutic potential. Angew Chem Int Ed 49:537–540

Bremner JB, Kelso MJ (2010b) Synthesis of berberine-efflux pump inhibitor hybrid antibacterials. Synth Commun 40 (23):3561–3568

Bremner JB (2017) Molecular design of potential triple-action antibacterial agents and related pro-drugs. SWU Sci J 33 (1):1–19

Brötz-Oesterhelt H, Brunner NA (2008) How many modes of action should an antibiotic have? Curr Opin Pharmacol 8 (5):564–573

Brown MF, Mitton-Fry MJ, Arcari JT et al. (2013) Pyridone-conjugated monobactam antibiotics with Gram-negative activity. J Med Chem 56 (13):5541–5552

Buege MJ, Brown JE, Aitken SL (2017) Solithromycin: A novel ketolide antibiotic. Am J Health-Syst Pharm 74(12):875–887

Caliceti C, Franco P, Spinozzi S et al. (2016) Berberine: New insights from pharmacological aspects to clinical evidences in the management of metabolic disorders. Curr Med Chem 23 (14):1460–1476

Casiraghi A, Suigo L, Valoti E et al. (2020) Targeting bacterial cell division: A binding site-centered approach to the most promising inhibitors of the essential protein FtsZ. Antibiotics 9 (2):69

CenterWatch (2019) FDA-approved drugs-CenterWatch; Infections and infectious diseases; Xenleta (lefamulin). https://www.centrewatch.com/drug-information/fda-approved-drugs/. Accessed 19 December 2020

Chan A, Shiver A, Wever W et al. (2017) Role for dithiolopyrrolones in disrupting bacterial metal homeostasis. Proc Natl Acad Sci USA 114 (10):2717–2722

Charrier C, Salisbury AM, Savage V J et al. (2017) Novel bacterial topoisomerase inhibitors with potent broad-spectrum activity against drug-resistant bacteria. Antimicrob Agents Chemother 61(5):e02100–16

Chen J, Wang T, Xu S et al. (2017) Design, synthesis and biological evaluation of novel nitric oxide-donating protoberberine derivatives as antitumor agents. Eur J Med Chem 132:173–183

Chen R, Zhang H, Ma T et al. (2019) Ciprofloxacin-1,2,3-triazole-isatin hybrids tethered via amide: Design, synthesis, and in vitro anti-mycobacterial activity evaluation. Bioorg Med Chem Lett 29 (18):2635–2637

Cheng J, Thanassi JA, Thoma CL et al. (2007) Dual targeting of DNA gyrase and topoisomerase IV: target interactions of heteroaryl isothiazolones in *Staphylococcus aureus*. Antimicrob Agents Chemother 51 (7):2445–2453

Cheng P, Wang B, Liu X et al. (2014) Facile synthesis of tetrahydroberberine and protoberberine alkaloids from protopines and study on their antibacterial activities. Nat Prod Res 28(7):413–419

Cheng Z, Chen A-F, Wu F et al. (2010) 8,8-Dimethyldihydroberberine with improved bioavailability and oral efficacy on obese and diabetic mouse models. Bioorg Med Chem 18:5915–5924

Colomer I, Empson CJ, Craven P et al. (2016) A divergent synthetic approach to diverse molecular scaffolds: assessment of lead-likeness using LLAMA, an open-access computational tool. ChemComm 52 (45):7209–7212

Costa SS, Viveiros M, Amaral L et al. (2013) Multidrug efflux pumps in *Staphylococcus aureus*: an update. Open Microbiol J 7:59–71

Dateer RB, Shaibu B, Liu R-S (2012) Gold-catalyzed intermolecular [4+2] cycloadditions of ynamides with alkenes. Angew Chem Int Ed 51:113–117

Dean CR, Barkan DT, Bermingham A et al. (2018) Mode of action of the monobactam LYS228 and mechanisms decreasing in vitro susceptibility in *Escherichia coli* and *Klebsiella pneumoniae*. Antimicrob Agents Chemother 62 (10):e01200–18

Decker M (2017) Introduction. In: Decker M (ed) Design of hybrid molecules for drug development. Elsevier, Amsterdam, pp 1–3

DeCorte B, Russcher JC, Monnee MCF (2018) Therapeutic compounds for pain and synthesis thereof. US Patent 10,040,795 B2, 7 Aug 2018

Decuyper L, Jukič M, Sosič I et al. (2018) Antibacterial and β-lactamase inhibitory activity of monocyclic β-lactams. Med Res Rev 38 (2):426–503

DeGoey DA, Chen HJ, Cox PB et al. (2018) Beyond the rule of 5: lessons learned from AbbVie's drugs and compound collection. J Med Chem 61 (7):2636–2651

Dolla N, Chen C, Larkins-Ford, J et al. (2014) On the mechanism of berberine-INF55 (5-nitro-2-phenylindole) hybrid antibacterials. Aust J Chem. 67:1471–1480

Domadia PN, Bhunia, A, Sivaraman J et al. (2008) Berberine targets assembly of *Escherichia coli* cell division protein FtsZ. Biochem 47(10): 3225–3234

Domalaon R, Idowu T, Zhanel GG et al. (2018) Antibiotic hybrids: the next generation of agents and adjuvants against Gram-negative pathogens? Clin Microbiol Rev 31 (2):e00077–00017

East SP, Silver LL (2013) Multitarget ligands in antibacterial research: progress and opportunities. Expert Opin Drug Discov 8 (2):143–156

Edwards MJ, Flatman RH, Mitchenall LA et al. (2009). A crystal structure of the bifunctional antibiotic simocyclinone D8, bound to DNA gyrase. Science 326:1415–1418

Eijkelkamp BA, McDevitt CA, Kitten T (2015) Manganese uptake and streptococcal virulence. Biometals 28 (3):491–508

Etukala JR, Kumar EVKS, Ablordeppey SY (2008) A short and convenient synthesis and evaluation of the antiinfective properties of indoloquinoline alkaloids: 10*H*-Indolo[3,2-b]quinoline and 7*H*-indolo[2,3-c]quinolines. J Het Chem 45 (2):507–512

Fan T, Guo X, Zeng Q et al. (2020) Synthesis and structure-activity relationship of palmatine derivatives as a novel class of antibacterial agents against *Helicobacter pylori*. Molecules 25:1352

Farha MA, Brown ED (2019) Drug repurposing for antimicrobial discovery. Nat Microbiol 4 (4):565–577

Fedorowicz J, Sączewski J (2018) Modifications of quinolones and fluoroquinolones: hybrid compounds and dual-action molecules. Monatsh Chem 149 (7):1199–1245

Foley DJ, Craven PGE, Collins PM et al. (2017) Synthesis and demonstration of the biological relevance of sp(3) -rich scaffolds distantly related to natural product frameworks. Chem Eur J 23 (60):15227–15232

Fujita K-i, Enoki Y, Yamaguchi R (2008) Cp*Ir-catalyzed N-alkylation of amines with alcohols. A versatile and atom economical method for the synthesis of amines. Tetrahedron 64:1943–1954

Gallardo-Godoy A, Hansford KA, Muldoon C et al. (2019) Structure-Function Studies of Polymyxin B Lipononapeptides. Molecules 24 (3):553

Gandini A, Bartolini M, Tedesco D et al. (2018) Tau-Centric multitarget approach for Alzheimer's disease: Development of first-in-class dual glycogen synthase kinase 3β and tau-aggregation inhibitors. J Med Chem 61 (17):7640–7656

Gharbo SA, Beal JL, Doskotch RW et al. (1973) Alkaloids of Thalictrum. XIV. Isolation of alkaloids having antimicrobial activity from *Thalictrum polygamum*. Lloydia 36 (3):349–351

Gorityala BK, Guchhait G, Goswami S et al. (2016) Hybrid antibiotic overcomes resistance in *P. aeruginosa* by enhancing outer membrane penetration and reducing efflux. J Med Chem 59 (18):8441–8455

Górska A, Sloderbach A, Marszałł MP (2014) Siderophore–drug complexes: potential medicinal applications of the 'Trojan horse' strategy. Trends Pharmacol Sci 35 (9):442–449

Grycová L, Dostál J, Marek R (2007) Quaternary protoberberine alkaloids. Phytochemistry 68 (2):150–175

Guerrera CA, Ryder TR (2016) The Petasis Borono-Mannich Multicomponent Reaction. In: Boron Reagents in Synthesis, vol 1236. ACS Symposium Series, vol 1236. American Chemical Society, pp 275–311

Guo D, Li J, Lin H et al. (2016) Design, synthesis, and biological evaluation of novel tetrahydroprotoberberine derivatives (THPBs) as selective α(1A)-adrenoceptor antagonists. J Med Chem 59 (20):9489–9502

Gupta V, Datta P (2019) Next-generation strategy for treating drug resistant bacteria: Antibiotic hybrids. Indian J Med Res 149 (2):97–106

Gupta V, Gaur R, Sharma A et al. (2018) A novel bi-functional chalcone inhibits multi-drug resistant *Staphylococcus aureus* and potentiates the activity of fluoroquinolones. Bioorg Chem 83:214–225

Hakobyan S, Boily J-F, Ramstedt M (2014) Proton and gallium(III) binding properties of a biologically active salicylidene acylhydrazide. J Inorg Biochem 138:9–15

Han S, Zaniewski RP, Marr ES et al. (2010) Structural basis for effectiveness of siderophore-conjugated monocarbams against clinically relevant strains of *Pseudomonas aeruginosa*. PNAS 107 (51):22002–22007

Harbut MB, Vilchèze C, Luo X et al. (2015) Auranofin exerts broad-spectrum bactericidal activities by targeting thiol-redox homeostasis. Proc Natl Acad Sci 112(14):4453–4458

Ho TT, Tran QT, Chai CL (2018) The polypharmacology of natural products. Future Med Chem 10 (11):1361–1368

Hopkins AL (2009) Drug discovery: Predicting promiscuity. Nature 462 (7270):167–168

Islas R, Chamorro E, Robles J et al. (2007) Borazine: To be or not to be aromatic. Struct Chem 18:833–839

Ito A, Nishikawa T, Matsumoto S et al. (2016) Siderophore cephalosporin Cefiderocol utilizes ferric iron transporter systems for antibacterial activity against *Pseudomonas aeruginosa*. Antimicrob Agents Chemother 60 (12):7396–7401

Ito A, Sato T, Ota M et al. (2018) In vitro antibacterial properties of Cefiderocol, a novel siderophore cephalosporin, against Gram-negative bacteria. Antimicrob Agents Chemother 62 (1):e01454–17

Ivanenkov YA,Yamidanov RS, Osterman IA et al. (2019a) 2-Pyrazol-1-yl-thiazole derivatives as highly potent antibacterials. J Antibiot 72:827–833

Ivanenkov YA, Zhavoronkov A, Yamidanov RS et al. (2019b) Identification of novel antibacterials using machine learning techniques. Front Pharmacol 10:913

Iwasa K, Nanba H, Lee DU et al. (1998) Structure-activity relationships of protoberberines having antimicrobial activity. Planta Med 64 (8):748–751

Janas A, Przybylski P (2019) 14- and 15-Membered lactone macrolides and their analogues and hybrids: structure, molecular mechanism of action and biological activity. Eur J Med Chem 182:111662

Jayaraman P, Sakharkar KR, Lim C et al. (2013) Novel phytochemical-antibiotic conjugates as multitarget inhibitors of *Pseudomonas aeruginosa* GyrB/ParE and DHFR. Drug Des Dev Ther 7: 449–475

Jaye JA, Gelinas BS, McCormick GM et al. (2017) Implications of the final ring closure to 10b-aza-10c-borapyrene for aryl–alkyne ring-closing mechanisms. Can J Chem 95 (4):357–362

Jin J-L, Hua G-q, Meng Z et al. (2010) Antibacterial mechanisms of berberine and reasons for little resistance of bacteria. Chin Herb Med 3(1):27–35
Kalia VC (2013) Quorum sensing inhibitors: An overview. Biotechnol Adv 31:224–245.
Kapp E, Malan SF, Joubert J et al. (2018) Small molecule efflux pump inhibitors in *Mycobacterium tuberculosis*: a rational drug design perspective. Mini Rev Med Chem 18 (1):72–86
Kaul G, Shukla M, Dasgupta A et al. (2019) Update on drug-repurposing: is it useful for tackling antimicrobial resistance? Future Microbiol 14:829–831
Kim SH, Lee SJ, Lee JH et al. (2002) Antimicrobial activity of 9-O-acyl- and 9-O-alkylberberrubine derivatives. Planta Med 68 (3):277–281
King TA, Stewart HL, Mortensen KT et al. (2019) Cycloaddition Strategies for the Synthesis of Diverse Heterocyclic Spirocycles for Fragment-Based Drug Discovery. Eur J Org Chem 2019 (31–32):5219–5229
Klahn P, Brönstrup M (2017) Bifunctional antimicrobial conjugates and hybrid antimicrobials. Nat Prod Rep 34 (7):832–885
Kohnen-Johannsen KL, Kayser O (2019) Tropane alkaloids: Chemistry, pharmacology, biosynthesis and production. Molecules 24(4):796
Koul A, Arnoult E, Lounis N et al. (2011) The challenge of new drug discovery for tuberculosis. Nature 469 (7331):483–490
Kourtesi C, Ball AR, Huang YY et al. (2013) Microbial efflux systems and inhibitors: approaches to drug discovery and the challenge of clinical implementation. Open Microbiol J 7(Suppl 1-M3):34–52
Krämer K (2019) A wish list for organic chemistry. Chem World 16 (4):34–35
Kusuma KD, Griffith R, Harry EJ et al. (2019) In silico analysis of ftsz crystal structures towards a new target for antibiotics. Aust J Chem 72:184–193
Kuznedelov K, Semenova E, Knappe TA et al. (2011) The antibacterial threaded-lasso peptide capistruin inhibits bacterial RNA polymerase. J Mol Biol 412 (5):842–848
Lamut A, Mašič L, Kikelj D et al. (2019) Efflux pump inhibitors of clinically relevant multidrug resistant bacteria. Med Res Rev 39:1–45
Lavrado J, Moreira R, Paulo A (2010) Indoloquinolines as scaffolds for drug discovery. Curr Med Chem 17 (22):2348–2370
Lessa JA, Parrilha GL, Beraldo H (2012) Gallium complexes as new promising metallodrug candidates. Inorganica Chim Acta 393:53–63
Li B, Wever WJ, Walsh CT et al. (2014) Dithiolopyrrolones: biosynthesis, synthesis, and activity of a unique class of disulfide-containing antibiotics. Nat Prod Rep 31 (7):905–923
Li R, Wu J, He Y et al. (2014a) Synthesis and in vitro evaluation of 12-(substituted aminomethyl)berberrubine derivatives as anti-diabetics. Bioorg Med Chem Lett 24:1762–1765
Li Y-H, Li Y, Yang P et al. (2010) Design, synthesis, and cholesterol-lowering efficacy for prodrugs of berberrubine. Bioorg Med Chem 18 (17):6422–6428
Lin Z, Xu X, Zhao S et al. (2018) Total synthesis and antimicrobial evaluation of natural albomycins against clinical pathogens. Nat Commun 9 (1):3445
Ling LL, Schneider T, Peoples AJ et al. (2015) A new antibiotic kills pathogens without detectable resistance. Nature 517:455–459
Liu Z, Li J, Li S et al. (2018) SuFEx click chemistry enabled late-stage drug functionalization. J Am Chem Soc 140:2919–2925
Livermore DM, Warner M, Mushtaq S et al. (2016). Interactions of OP0595, a novel triple-action diazabicyclooctane, with β-lactams against OP0595-resistant *Enterobacteriaceae* mutants. Antimicrob Agents Chemother 60(1): 554–560
Livermore DM, Mushtaq, S, Warner M et al. (2017) In vitro activity of cefepime/zidebactam (WCK 5222) against Gram-negative bacteria. J Antimicrob Chemother 72(5):1373–1385
Lo C-Y, Hsu L-C, Chen M-S et al. (2013) Synthesis and anti-cancer activity of a novel series of 9-*O*-substituted berberine derivatives: A lipophilic substitute role. Bioorg Med Chem Lett 23: 305–309

Lovering F, Bikker J, Humblet C (2009) Escape from flatland: Increasing saturation as an approach to improving clinical success. J Med Chem 52 (21):6752–6756

Lowe RA, Taylor D, Chibale K et al. (2020) Synthesis and evaluation of the performance of a small molecule library based on diverse tropane-related scaffolds. Bioorg Med Chem 28 (9):115442

Ma Z, Lynch AS (2016) Development of a dual-acting antibacterial agent (TNP-2092) for the treatment of persistent bacterial infections. J Med Chem 59 (14):6645–6657

Mahjour B, Shen Y, Liu W et al. (2020) A map of the amine–carboxylic acid coupling system. Nature 580 (7801):71–75

Mahmood HY, Jamshidi S, Sutton JM et al. (2016) Current advances in developing inhibitors of bacterial multidrug efflux pumps. Curr Med Chem 23 (10):1062–1081

Maianti JP, Hanessian S (2016) Structural hybridization of three aminoglycoside antibiotics yields a potent broad-spectrum bactericide that eludes bacterial resistance enzymes. Med Chem Commun 7:170–176

Mari G, De Crescentini L, Benedetti S et al. (2020) Synthesis of new dihydroberberine and tetrahydroberberine analogues and evaluation of their antiproliferative activity on NCI-H1975 cells. Beilstein J Org Chem 16:1606–1616

Martin JK, 2nd, Sheehan JP, Bratton BP et al. (2020) A dual-mechanism antibiotic kills Gram-negative bacteria and avoids drug resistance. Cell 181 (7):1518–1532

Meanwell NA (2011) Synopsis of some recent tactical application of bioisosteres in drug design. J Med Chem 54 (8):2529–2591

Mehedi MAU, Molla A, Khondkar P et al. (2010) Pseurotin A: An antibacterial secondary metabolite from *Aspergillus fumigatus*. Asian J Chem 22:2611–2614

Millan M (2009) Dual- and triple-acting agents for treating core and co-morbid symptoms of major depression: novel concepts, new drugs. Neurotherapeutics 6 (1):53–77

Miller M, Zhu H, Xu Y et al. (2009) Utilization of microbial iron assimilation processes for the development of new antibiotics and inspiration for the design of new anticancer agents. Biometals 22:61–75

Mishra SK, Tripathi G, Kishore N et al. (2017) Drug development against tuberculosis: Impact of alkaloids. Eur J Med Chem 137: 504–544

Mistry B, Patel RV, Keum Y-S et al. (2017) Synthesis of *N*-Mannich bases of linking piperazine moieties revealing anticancer and antioxidant effects. Saudi J Biol Sci 24:36–44

Monteleone S, Fuchs JE, Liedl KR (2017) Molecular connectivity predefines polypharmacology: Aliphatic rings, chirality, and sp^3 centers enhance target selectivity. Front Pharmacol 8:552

Morgentin R, Dow M, Aimon A et al. (2018) Translation of innovative chemistry into screening libraries: an exemplar partnership from the European Lead Factory. Drug Discov Today 23 (9):1578–1583

Morinaka A, Tsutsumi Y, Yamada M et al. 2015. OP0595, a new diazabicyclooctane: Mode of action as a serine β-lactamase inhibitor, antibiotic and β-lactam 'enhancer'. J Antimicrob Chemother 70(10): 2779–2786

Mori-Quiroz LM, Hedrick SL, De Los Santos AR et al. (2018) A unified strategy for the syntheses of the isoquinolinium alkaloids berberine, coptisine, and jatrorrhizine. Org Lett 20 (14):4281–4284

Morphy R, Rankovic Z (2006) The physicochemical challenges of designing multiple ligands. J Med Chem 49 (16):4961–4970

Mosaei H, Molodtsov V, Kepplinger B et al. (2018) Mode of action of kanglemycin A, an ansamycin natural product that is active against rifampicin-resistant *Mycobacterium tuberculosis*. Mol Cell 72:263–274

Murphy AC, Gao S-S, Han L-C et al. (2014) Biosynthesis of thiomarinol A and related metabolites of *Pseudoalteromonas* sp. SANK 73390. Chem Sci 5 (1):397–402

Nagao Y (2013) Chemical pharma-sciences that incorporate non-covalent bonded interactions. Heterocycles 87 (1): 1–29

Nagao Y, Ilmori H, Nam KH et al. (2001) Synthesis and antibacterial activity of new 1b-Methylcarbapenems having the potential for intramolecular nonbonded S···O interactions. Chem Pharm Bull 49 (12):1660–1661

Naruto S, Mizuta H, Nishimura H (1976) A novel amination of isoquinolinium salts via nucleophilic substitution reaction. Tetrahedron Lett 17 (19):1597–1600

Nasr T, Bondock S, Ibrahim TM et al. (2020) New acrylamide-sulfisoxazole conjugates as dihydropteroate synthase inhibitors. Bioorg Med Chem 28 (9):115444

Nechepurenko IV, Salakhutdinov NF, Tolstikov GA (2010) Berberine chemistry and biological activity. Chem Sustain Dev 18:1–23

Okano A, Isley N, Boger D (2017) Peripheral modifications of [Ψ[CH 2 NH]Tpg 4] vancomycin with added synergistic mechanisms of action provide durable and potent antibiotics. Proc Natl Acad Sci U.S.A. 114:201704125

Oldfield E, Feng X (2014) Resistance-resistant antibiotics. Trends Pharmacol Sci 35 (12):664–674

Olleik H, Yacoub T, Hoffer L et al. (2020) Synthesis and evaluation of the antibacterial activities of 13-substituted berberine derivatives. Antibiotics 9 (7):381

Opperman TJ, Nguyen ST (2015) Recent advances toward a molecular mechanism of efflux pump inhibition. Front Microbiol 6: Article 421

Owens B (2017) Solithromycin rejection chills antibiotic sector. Nat Biotechnol 35 (3):187–188

Ozçelik B, Kartal M, Orhan I (2011) Cytotoxicity, antiviral and antimicrobial activities of alkaloids, flavonoids, and phenolic acids. Pharm Biol 49 (4):396–402

Park KD, Lee JH, Kim SH et al. (2006) Synthesis of 13-(substituted benzyl) berberine and berberrubine derivatives as antifungal agents. Bioorg Med Chem Lett 16 (15):3913–3916

Parker EN, Drown BS, Geddes EJ et al. (2020) Implementation of permeation rules leads to a FabI inhibitor with activity against Gram-negative pathogens. Nat Microbiol 5 (1):67–75

Parkes AL, Yule IA (2016) Hybrid antibiotics - clinical progress and novel designs. Expert Opin Drug Discov 11 (7):665–680

Parmar A, Iyer A, Lloyd DG et al. (2017) Syntheses of potent teixobactin analogues against methicillin-resistant *Staphylococcus aureus* (MRSA) through the replacement of l-allo-enduracididine with its isosteres. Chem Commun 53 (55):7788–7791

Peek J, Lilic M, Montiel D et al. (2018) Rifamycin congeners kanglemycins are active against rifampicin-resistant bacteria via a distinct mechanism. Nat Commun 9 (1):4147

Peek J, Xu J, Wang H et al. (2020) A semisynthetic Kanglemycin shows in vivo efficacy against high-burden rifampicin resistant pathogens. ACS Infect Dis 6:2431–2440

Peterson RM, Huang T, Rudolf JD et al. (2014) Mechanisms of self-resistance in the platensimycin- and platencin-producing *Streptomyces platensis* MA7327 and MA7339 strains. Chem Biol 21 (3):389–397

Pitts CR, Bornemann D, Liebing P et al. (2019) Making the SF_5 group more accessible: A gas-reagent-free approach to aryl tetrafluoro-λ^6-sulfanyl chlorides. Angew Chem Int Ed 58:1950–1954

Pokrovskaya V, Baasov T (2010) Dual-acting hybrid antibiotics: a promising strategy to combat bacterial resistance. Expert Opin Drug Discov 5 (9):883–902

Rajendran L, Knölker HJ, Simons K (2010) Subcellular targeting strategies for drug design and delivery. Nat Rev Drug Discov 9 (1):29–42

Ramesh R, Reddy DS (2018) Quest for novel chemical entities through incorporation of silicon in drug scaffolds. J Med Chem 61 (9):3779–3798

Richter MF, Drown BS, Riley AP et al. (2017) Predictive compound accumulation rules yield a broad-spectrum antibiotic. Nature 545 (7654):299–304

Roselli M, Cavalluzzi MM, Bruno C et al. (2016) Synthesis and evaluation of berberine derivatives and analogs as potential antiacetylcholinesterase and antioxidant agents. Phytochem Lett 18:150–156

Rzhepishevska O, Ekstrand-Hammarström B, Popp M et al. (2011) The antibacterial activity of Ga3+ is influenced by ligand complexation as well as the bacterial carbon source. Antimicrob Agents Chemother 55 (12):5568–5580

Sader HS, Rhomberg PR, Flamm RK et al. (2017) WCK 5222 (cefepime/zidebactam) antimicrobial activity tested against Gram-negative organisms producing clinically relevant ß-lactamases. J Antimicrob Chemother 72(6):1696–1703

Samosorn S, Tanwirat B, Muhamad N et al. (2009) Antibacterial activity of berberine-NorA pump inhibitor hybrids with a methylene ether linking group. Bioorg Med Chem 17 (11):3866–3872

Schindler B, Jacinto P, Kaatz G (2013) Inhibition of drug efflux pumps in *Staphylococcus aureus*: Current status of potentiating existing antibiotics. Future Microbiol 8:491–507

Schwarz B, Merkel OM (2019) Nose-to-brain delivery of biologics. Ther Deliv 10 (4):207–210

Sebastiano MR, Doak BC, Backlund M et al. (2018) Impact of dynamically exposed polarity on permeability and solubility of chameleonic drugs beyond the rule of 5. J Med Chem 61:4189–4202

Shapiro, S. 2013. Speculative strategies for new antibacterials: All roads should not lead to Rome. J Antibiot 66: 371–386

Shavit M, Pokrovskaya V, Belakhov V et al. (2017) Covalently-linked kanamycin-ciprofloxacin hybrid antibiotics as a tool to fight bacterial resistance. Bioorg Med Chem 25(11): 2917–2925

Shukla R, Medeiros-Silva J, Parmar A et al. (2020) Mode of action of teixobactins in cellular membranes. Nat Commun 11 (1):2848

Silver LL (2007) Multi-targeting by monotherapeutic antibacterials. Nat Rev Drug Discov 6 (1):41–55

Simoni E, Daniele S, Bottegoni G et al. (2012) Combining galantamine and memantine in multi-targeted, new chemical entities potentially useful in Alzheimer's disease. J Med Chem 55 (22):9708–9721

Sowaileh MF, Hazlitt RA, Colby DA (2017) Application of the pentafluorosulfanyl group as a bioisosteric replacement. ChemMedChem 12: 1481–490

Stokes N, Baker N, Bennett J et al. (2013) An improved small-molecule inhibitor of FtsZ with superior in vitro potency, drug-like properties, and in vivo efficacy. Antimicrob Agents Chemother 57 (1):317–325

Stokes JM, MacNair CR, Ilyas B et al. (2017) Pentamidine sensitizes Gram-negative pathogens to antibiotics and overcomes acquired colistin resistance. Nat Microbiol 2: 17028

Stoodley RJ (1977) Aspects of heterocyclic synthesis. Chem Ind 8 (39):377–385

Sun N, Chan F-Y, Lu Y-J et al. (2014) Rational design of berberine-based FtsZ inhibitors with broad-spectrum antibacterial activity. PloS one 9 (5):e97514–e97514

Sun T, Li X-D, Hong J et al. (2019) Inhibitory effect of two traditional chinese medicine monomers, Berberine and Matrine, on the quorum sensing system of antimicrobial-resistant *Escherichia coli*. Front Microbiol 10:2584–2584

Tague AJ, Putsathit P, Hutton ML et al. (2019a) Cationic biaryl 1,2,3-triazolyl peptidomimetic amphiphiles targeting *Clostridioides* (*Clostridium*) *difficile*: Synthesis, antibacterial evaluation and an in vivo *C. difficile* infection model. Eur J Med Chem 170:203–224

Tague AJ, Putsathit P, Hammer KA et al. (2019b) Cationic biaryl 1,2,3-triazolyl peptidomimetic amphiphiles: synthesis, antibacterial evaluation and preliminary mechanism of action studies. Eur J Chem 168:386–404

Tahir S, Mahmood T, Dastgir F et al. (2019) Design, synthesis and anti-bacterial studies of piperazine derivatives against drug resistant bacteria. Eur J Med Chem 166:224–231

Tambat R, Jangra M, Mahey N et al. (2019) Microbe-derived indole metabolite demonstrates potent multidrug efflux pump inhibition in *Staphylococcus aureus*. Front Microbiol 10:2153–2153

Tang X, Zhang J, Liu L et al. (2017) Synthesis of 13-substituted derivatives of berberine: Aggregation-induced emission enhancement and pH sensitive property. J Photochem Photobiol A 337:71–81

Teng Q, Zhu X, Guo Q et al. (2019) Synthesis of 9-*O*-arylated berberines via copper-catalyzed C(Ar)-O coupling reactions. Beilstein J Org Chem 15:1575–1580

Tevyashova A, Olsuf'eva E, Preobrazhenskaya M (2015) Design of dual action antibiotics as an approach to search for new promising drugs. Russ Chem Rev 84:61–97

Thangamani S, Mohammad H, Abushahba MFN et al. (2016) Antibacterial activity and mechanism of action of auranofin against multi-drug resistant bacterial pathogens. Sci. Rep 6:22571

Thirumurugan P, Matosiuk D, Jozwiak K (2013) Click chemistry for drug development and diverse chemical-biology applications. Chem Rev 113(7): 4905–4979

Tomkiewicz D, Casadei G, Larkins-Ford J et al. (2010) Berberine-INF55 (5-nitro-2-phenylindole) hybrid antimicrobials: Effects of varying the relative orientation of the berberine and INF55 components. Antimicrob Agents Chemother 54:3219–3224

Tumey LN, Boschelli DH, Bhagirath N et al. (2014) Identification and optimization of indolo[2,3-c]quinoline inhibitors of IRAK4. Bioorg Med Chem Lett 24 (9):2066–2072

Van Bambeke F, Pagès J-M, Lee V (2009) Inhibitors of bacterial efflux pumps as adjuvants in antibacterial therapy and diagnostic tools for detection of resistance by efflux. In: Atta-ur-Rahman, Choudhary MI (eds) Frontiers in anti-infective drug discovery, vol 1. Bentham Science, pp 138–175

Van der Schyf CJ, Geldenhuys WJ (2009) Polycyclic compounds: ideal drug scaffolds for the design of multiple mechanism drugs? Neurotherapeutics 6 (1):175–186

Van Miert S, Hostyn S, Maes BUW et al. (2005) Isoneocryptolepine, a synthetic indoloquinoline alkaloid, as an antiplasmodial lead compound. J Nat Prod 68:674–677

Wang B, Deng A-J, Li Z-H et al. (2020) Syntheses and structure–activity relationships in growth inhibition activity against human cancer cell lines of 12 substituted berberine derivatives. Molecules 25 (8):1871

Wang J, Kodali S, Lee SH et al. (2007) Discovery of platencin, a dual FabF and FabH inhibitor with in vivo antibiotic properties. Proc Natl Acad Sci 104 (18):7612–7616

Wang J, Tang J, Yu L-F et al. (2015) Berberine analogues: progress towards versatile applications. Heterocycles 91:2233

Wang J, Yang T, Chen H et al. (2017a) The synthesis and antistaphylococcal activity of 9, 13-disubstituted berberine derivatives. Eur J Med Chem 127:424–433

Wang K, Feng X, Chai L et al. (2017b) The metabolism of berberine and its contribution to the pharmacological effects. Drug Metab Rev 49 (2):139–157

Wang YX, Pang WQ, Zeng QX et al. (2018) Synthesis and biological evaluation of new berberine derivatives as cancer immunotherapy agents through targeting IDO1. Eur J Med Chem 143:1858–1868

Whitty A, Zhong M, Viarengo L et al. (2016) Quantifying the chameleonic properties of macrocycles and other high-molecular-weight drugs. Drug Disc Today 21(5):712–717

Wockhardt Discovery (2017) WCK 2349 (Oral prodrug). https://wockhardtdiscovery.com/wck-2349-oral-prodrug/. Accessed 6 March 2017

Woods A, Rahman KM (2018) Antimicrobial molecules in the lung: formulation challenges and future directions for innovation. Future Med Chem 10 (5):575–604

Xiao ZP, Wang XD, Wang PF et al. (2014) Design, synthesis, and evaluation of novel fluoroquinolone-flavonoid hybrids as potent antibiotics against drug-resistant microorganisms. Eur J Med Chem 80:92–100

Xue M, Yang Y, Chi X et al. (2015) Development of pseodorotaxanes and rotaxanes: From synthesis to stimuli-responsive motions to applications. Chem Rev 115:7398–7501

Yang Y-S, Wei W, Hu X-X et al. (2019) Evolution and antibacterial evaluation of 8-hydroxy-cycloberberine derivatives as a novel family of antibacterial agents against MRSA. Molecules 24 (5):984

Yi S, Varun BV, Choi Y et al. (2018) A brief overview of two major strategies in diversity-oriented synthesis: build/couple/pair and ring-distortion. Front Chem 6:507–507

Yin Z, Whittell LR, Wang Y et al. (2014) Discovery of lead compounds targeting the bacterial sliding clamp using a fragment-based approach. J Med Chem 57(6):2799–2806

Yin Z, Whittell LR, Wang Y et al. (2015) Bacterial sliding clamp inhibitors that mimic the sequential binding mechanism of endogenous linear motifs. J Med Chem 58 (11):4693–4702

Yuan Y, Pei J, Lai L (2020a) LigBuilder V3: a multi-target de novo drug design approach. Front Chem 8:142

Yuan Y, Wang X, Xu X et al. (2020b) Evaluation of a dual-acting antibacterial agent, TNP-2092, on gut microbiota and potential application in the treatment of gastrointestinal and liver disorders. ACS Infect Dis 6 (5):820–831

Zajdel P, Kos T, Marciniec K et al. (2018) Novel multi-target azinesulfonamides of cyclic amine derivatives as potential antipsychotics with pro-social and pro-cognitive effects. Eur J Med Chem 145:790–804

Zhang G-B, Maddili SK, Tangadanchu VKR et al. (2018) Discovery of natural berberine-derived nitroimidazoles as potentially multi-targeting agents against drug-resistant *Escherichia coli*. Sci China Chem 61 (5):557–568

Zhang W, Zhang D, Stashko MA et al. (2013) Pseudo-cyclization through intramolecular hydrogen bond enables discovery of pyridine substituted pyrimidines as new Mer Kinase inhibitors. J Med Chem 56 (23):9683–9692

Zheng Y, Tice CM, Singh SB (2014) The use of spirocyclic scaffolds in drug discovery. Bioorg Med Chem Lett 24:3673–3682

Zhang Y, Shi W, Zhang W et al. (2014) Mechanisms of pyrazinamide action and resistance. Microbiol Spectr 2 (4):MGM2-0023-2013

Zhou S, Tong R (2016) A general, concise strategy that enables collective total syntheses of over 50 protoberberine and five aporhoeadane alkaloids within four to eight steps. Chem Eur J 22 (21):7084–7089

Zhou X, Chen M, Zheng Z et al. (2017) Synthesis and evaluation of novel 12-aryl berberine analogues with hypoxia-inducible factor-1 inhibitory activity. RSC Adv 7:26921–26929

Zhu R, Wang Y, Liu J et al. (2017) One-pot synthesis of imidazolyl isoquinolines under a palladium-catalyzed C-H activation/annulation (CHAA) reaction. Synthesis 49:1335–1341

Zimmermann S, Klinger-Strobel M, Bohnert JA et al. (2019) Clinically approved drugs inhibit the *Staphylococcus aureus* multidrug NorA efflux pump and reduce biofilm formation. Front Microbiol 10:2762–2762

Zong Y, Fang F, Meyer KJ et al. (2019) Gram-scale total synthesis of teixobactin promoting binding mode study and discovery of more potent antibiotics. Nat Commun 10 (1):3268

Chapter 4
Design Principles and Development of Prodrugs for Multiply Active Antibacterials

"Every problem is an opportunity in disguise."
—John Adams, American Statesman and Second U.S. President.

Abstract Added layers of complexity in design are inherent with multi-action agents released from prodrugs, and though challenging, such prodrugs do offer potential advantages over the non-cleavable multi-action hybrids. In particular the pro-moiety of the prodrug can be designed to increase solubility and permeability while the prodrug itself can enable effective delivery and site specific release of the bioactive component or components. This is particularly the case if the prodrug can be concentrated in bacteria or on their surface or in close proximity to them. The shorthand terminology 'multiply active prodrugs' covers compounds which need to be cleaved or activated to afford a multiply active product, or products, leading to antibacterial action. This chapter discusses known prodrugs affording multiply active products on cleavage and expands the conceptual framework to new possibilities for the design of such prodrugs. A number of release mechanisms are considered (enzymatic, chemical and physical) and implications for bacterial selectivity are explored in this Chapter.

J. Bremner, *Multiple Action-Based Design Approaches to Antibacterials*,
https://doi.org/10.1007/978-981-16-0999-2_4

4.1 Definitions

The generally accepted definition of a **prodrug** is the one specified in the glossary of terms used in medicinal chemistry by Wermuth et al. (1998), as follows:

"A prodrug is any compound that undergoes biotransformation before exhibiting its pharmacological effects. Prodrugs can also be viewed as drugs containing specialized non-toxic protective groups used in a transient manner to alter or to eliminate undesirable properties in the parent molecule."

Thus, according to this definition, a prodrug has no inherent pharmacological activity but is convertible to a compound or compounds that do. The biotransformations themselves may involve chemical conversion via enzymatic or non-enzymatic processes.

Prodrugs are then normally classified into two general sub-groups, one designated as **carrier-linked prodrugs (carrier prodrugs)** and the other as **bioprecursor prodrugs**. While a more recent alternative general classification has been proposed based on whether the bioactivation occurs intracellularly or extracellularly (Zawilska et al. 2013), the earlier classification, together with the sub-groups, is referred to in this book as they are considered more useful from a medicinal chemistry design viewpoint.

4.1.1 Carrier-Linked Prodrugs (Carrier Prodrugs)

Further to the IUPAC glossary of terms, "a carrier-linked prodrug (carrier prodrug) is a prodrug that contains a temporary linkage to a given active substance with a transient carrier group that produces improved physicochemical or pharmacokinetic properties and that can be easily removed in vivo, usually by a hydrolytic cleavage." (Wermuth et al. 1998).

Whilst this is a clear general definition, researchers in the field have found it helpful from the point of view of design parameters to differentiate further sub-divisions of the carrier prodrug category. Although this adds some complications, such further systemisation highlights gaps and supports pattern recognition endeavours. Carrier-linked prodrugs or carrier prodrugs can be designed in two parts (bipartite) composed of one carrier group or promoiety directly attached to the drug or in three parts (tripartite). In the latter case the carrier group is attached via a linker to the drug with the linker being the extra part. Further delineation of the carrier prodrug category (Wermuth et al. 1998; Parajuli et al. 2015; Abet et al. 2017) includes the following identified sub-groups:

(i) a **double prodrug**, that is "a biologically inactive molecule which is transformed in vivo in two steps (enzymatically and/or chemically) to the active species." (Wermuth et al. 1998). This type of prodrug is also referred to as a 'pro-prodrug' but it should perhaps be named, preferably, as a 'pre-prodrug'. This pro-prodrug concept, which was proposed by Bundgaard in

1989 (Bundgaard 1989), has also been classified separately as a third main group (Jubeh et al. 2020) but not always. An example of a double prodrug is that of pivampicillin, a double ester, lipid-soluble prodrug of the antibiotic ampicillin. Esterase-mediated hydrolysis of the terminal ester group in this prodrug then exposes a chemically unstable hydroxymethyl ester goup which fragments to ampicillin plus formaldehyde. This prodrug type can be extended to a triple prodrug design, in which three steps would be involved in the transformation to the active drug.

(ii) a **cascade prodrug** which is related to a double prodrug, but is defined separately in the IUPAC glossary of terms as being one "for which the cleavage of the carrier group becomes effective only after unmasking an activating group." (Wermuth et al. 1998).
Cascade prodrugs are normally either bipartite in design with the drug directly attached to a procarrier or in the tripartite category with a drug-linker-procarrier design.

(iii) a **mutual prodrug** which "is the association in a unique molecule of two, usually synergistic, drugs attached to each other, one drug being the carrier for the other and vice versa" (Wermuth et al. 1998). With a mutual prodrug, or also referred to as a co-drug (Das et al. 2010), the two synergistic drugs are chemically linked, with a view to improving the delivery properties of one or both drugs. The drugs treat the same disease but may mediate treatment via different mechanisms of action after selective cleavage in vivo. An example here is the orally administered drug, Sultamicillin, in which ampicillin is joined to the irreversible β-lactamase inhibitor sulfabactam via a methylene linked diester derived from the carboxylic acid groups on each. After absorption, esterase-mediated ester hydrolysis occurs followed by spontaneous formaldehyde loss resulting in self-immolation of the linker (Rautio et al. 2018) and release of the broad-spectrum antibiotic ampicillin plus sulbactam (Singh 2004).

(iv) a **macromolecular prodrug** which is one in which the carriers used involve macromolecules including polysaccharides, dextrans, cyclodextrins, peptides, proteins and other polymers (Parajuli et al. 2015).

(v) a **site-specific prodrug** in which the prodrug is directed to a specific site for release. This can be particularly important to avoid unwanted off-target actions and to enhance efficacy. With a site-specific prodrug the carrier can play a salient targeting role (Parajuli et al. 2015).

4.1.2 Bioprecursor Prodrugs

According to Wermuth et al. (1998) a "bioprecursor prodrug is a prodrug that does not imply the linkage to a carrier group, but results from a molecular modification of the active principle itself. This modification generates a new compound, able to be transformed metabolically or chemically, to the resulting compound being the active principle." Thus with these prodrugs there is no separate carrier group and no linkage. The molecular modification normally involves some functional group change but it can be a fine distinction between a carrier group and a directly attached functional group unit susceptible to enzymatic or non-enzymatic change.

Bioprecursor prodrugs can also be targeted through site-specific activation where the modifying functional group is activated by a specific enzyme at the site. Illustrative of this approach is the prodrug Secnidazole. Secnidazole is a marketed 5-nitroimidazole-based bactericidal drug for the treatment in adult women of bacterial vaginosis. Secnidazole is thought to enter bacteria through passive diffusion and the nitro group is then reduced by bacterial nitroreductases within the bacterium with the active radical anion intermediate produced then inhibiting DNA synthesis (Subbaiah and Meanwell 2019; Rautio et al. 2018). These useful reviews also cover prodrug design approaches to improve formulations as well as pharmacokinetic and targeting properties with a range of drugs.

4.2 Introduction to Prodrugs for Triple or Higher Action Antibacterials

4.2.1 Design Considerations

As with the design of hybrids, with prodrug design, for both carrier prodrugs and bioprecursor prodrugs, the overall goal should be to incorporate the minimum number of atoms or groups to achieve a number of sub-goals including: as low a molecular weight as possible, atom arrangements in the bioactive component to minimise or avoid unwanted stereoelectronic issues on interaction at the target sites, good pharmacokinetic properties after oral absorption, bacterial targeting and selective release, potent antibacterial activity for resistant and non-resistant pathogens, and maintenance of activity after likely or predicted changes in the pathogen targets to overcome the antibacterial. If possible every atom or group should be cut back to a minimum while maintaining efficacy in the active drug(s) released. With prodrugs, extra atoms are necessarily required in the promoiety to meet the penetration and selectivity requirements, but one should also consider further reinforcing functions for this moiety once released from the drug. These prodrug design paradigms and challenges are also covered in the excellent review by Rautio and co-authors (Rautio et al. 2018).

A further important consideration with any prodrug is to achieve good targeting to minimise unwanted side-effects; good molecular 'targeting' once the active drug is exposed is also vital. In general the initial targeting is realised through either site-directed delivery or site-specific bioactivation (Rautio et al. 2018). Enzymes are normally utilised to make or break bonds or undertake specific functional group transformations to create the active drug on or near the site of the target interactions. With antibacterials both site-directed and site-specific activation approaches have been assessed although the latter is more common. Multiply active prodrug design needs to consider both of these general specificity approaches. Controlling the interplay between selectivity, release mechanisms and nature of the released compounds is important in the successful antibacterial application of prodrugs.

Sometimes useful general pointers towards achieving selective targeting of a prodrug in one area can be obtained by looking at approaches used for tackling these issues in other therapies. For example methods used for ligand targeting for cancer (Rajendran et al. 2010; Srinivasarao et al. 2015) could be very useful in terms of targeting antibacterial prodrugs with nuanced linker design and release mechanisms. The design of the anti-cancer prodrugs discussed by Srinivasarao et al. (2015) is based on a 'targeting ligand-spacer-cleavable bridge-warhead' molecular structure in which the targeting ligand component is crucial for site-directed delivery using the ligand interaction with upregulated receptors on the cancer cell surface. Following binding and endocytosis a number of bridge cleavage processes may occur, depending on the nature of the linkage present, such as acid-catalysed hydrolysis in the acidic lysosome or enzymatic cleavage of a peptide linker by lysosomal cathepsin B, or by reduction of a disulfide linker by excess intracellular glutathione triggering subsequent intramolecular displacement of the active drug. From the antibacterial point of view, incorporation of a cleavable disulfide bridge would be relevant as there is no equivalent of endosomes/lysosomes in the bacterial cell. Also simplification of the design would be desirable and might be possible based on a disulfide linker but a suitable bacterial targeting ligand would still be vital. Glutathione plays important roles in bacteria in maintaining appropriate cytoplasmic redox balance and through blocking toxic compounds, amongst other roles. Bacteria have high concentrations of GSH, and GSH transporters have been identified in bacteria.

Overall, from a prodrug multi-targeting design perspective, one would need to consider a cleavable single compound which may be converted selectively in vivo to one, two or more active compounds. These product compounds may each have a single action or one or more of them may have multiple actions through other target site interactions. One of the products might, for example, be a dual action hybrid. Sometimes referred to as hybrid prodrugs, such hybrid active compound release after enzymatic activation is well known in the literature (Domalaon et al. 2018). In another expression of this, one active compound may be released as a triple action hybridic compound. For such potential triple action prodrugs similar general structural motifs can be envisaged as noted for the triple action hybrid designs in Chap. 3 (Sect. 3.3) but with the important incorporation of a cleavable group or groups. In order to reduce potential negative off target effects in vivo, the cleavage reaction should be triggered only when near or within the bacterium. Selectivity is important here, and such triggers could include bacterially specific enzymes or non-enzymatic mechanisms after selective accumulation.

4.2.2 Classification and Examples of Cleavable Types for Triple or Higher Action

With regard to these general type classifications, the letter descriptors A′, B′, or in some cases C′, are used for changes following prodrug activation and bond cleavage in order to take into account residual atoms still attached but retaining the required target recognition features. It should also be noted that for each cleavable type there could be variations in the sequencing of A, B, and C and the cleavage of the linking groups.

4.2.3 Cleavable Type I

The first cleavable type can be envisaged here as:

$$\mathbf{A\text{- - -}B \rightarrow A'\text{- - -}B \rightarrow A'' + B \quad or \quad A\text{- - -}B \rightarrow A + B'}$$

One hypothetical manifestation of this type could, in the first step for example, involve light activation of a photosensitiser moiety A, for example with cyanine groups to absorb light in the near-IR region (650–900 nm) to give A* which could then afford singlet oxygen from triplet oxygen and return to the ground state A. Subsequent reaction of singlet oxygen with the cyanine could then give a new intermediate A′ attached to B which is then susceptible to cleavage by chemical means in one or more steps to ultimately release the multiply active antibacterial B plus A″ (Nani et al. 2015). Nani and co-workers have used this packaged strategy combined with an attached antibody for cancer cell targeting. While somewhat complicated in design, such a strategy could be adapted in the antibacterial area to the selective release of the berberine derivative, berberrubine, with its phenolic group characteristics at C9 (Li et al. 2010). The effectiveness of such a strategy would hinge on a number of factors including selective accumulation on or in bacterial cells over host cells. An added benefit may be further singlet oxygen-mediated side reactivity although if these reactions predominated then the photooxidation of the cyanine portion would be reduced.

Using light, one could also envisage ultimate cleavage of A---B to give A plus B′ units via the intermediacy of singlet oxygen and a cleavable aminoacrylate linker analogous to the release of the anti-cancer drug Paclitaxel initiated by far red light activation (Thapa et al. 2016). Possibly this could be adapted to visible light activated prodrug design for antibacterials in the context of photodynamic therapy and in association with antibiotic (or modified antibiotic) release in situ to interact synergistically with an intracellular target or possibly a cell wall target. The antibiotic might be connected to the amino acrylate via an ester linkage using a hydroxyl or phenolic group on the antibiotic for this ester formation. Subsequent to singlet oxygen formation and cleavage of the amino acrylate unit (Bio et al. 2012), ready hydrolysis of the aldehyde phenolic or non-phenolic ester should occur to release the antibiotic. In the anti-cancer work by the same group and discussed by Thapa et al. (2016) a secondary alcoholic unit in Paclitaxel was used to link the aminoacrylate unit and the drug was released after ready hydrolysis of the aldehyde ester resulting from cleavage of the aldehyde ester intermediate.

Taking this particular prodrug approach a step further one might suggest a reversed photosensitiser-ciprofloxacin light activated system here linked by an aminoacrylamide unit with an appropriate spacer to the photosensitiser. With such a prodrug construct one could then expect multi-targeting through a combined release of an antibacterial drug together with the singlet oxygen (and other reactive oxygen species) produced, although some singlet oxygen would also be consumed in reacting with the enamide linking unit. If ciprofloxacin was connected via its piperazine secondary amino group directly, the other main product, an *N*-formyl piperidine might be structurally tuned to be a peptide deformylase (PDF) enzyme inhibitor to add to the ultimate activity profile. The other cleavage product would still retain the photosensitiser moiety for continuing the separate singlet oxygen production.

Other proposed examples in the type i category might include a structure A---B which is initially inactive but where A is susceptible to reversible ring opening of a β-lactam moiety by a lactamase to give an intermediate A′---B which could then chemically release B′ and ultimately re-form an active lactam antibacterial A″. If B′ was another enzyme inhibitor, for example of the enzyme peptide deformylase and A″ bound to two PBPs, then this would be a triple mode of action. The peptide deformylase inhibitor could be released by a β-lactamase enzyme or PBP from a suitably substituted cephalosporin and thus affecting protein synthesis as well as cell wall biosynthesis. The design of a suitable PDF inhibitor should be greatly assisted by knowledge of the binding modes of distinct PDF inhibitors as elaborated through the PDF platform discussed by Fieulaine and co-workers (Fieulaine et al. 2016). One could perhaps profitably incorporate the naturally-occurring PDF inhibitor, actinonin (Fig. 4.1a), in the prodrug design for ultimate release after a β-lactam cleavage initiation process. Linkage of the cephalosporin unit via the hydroxamic acid group, which is also present in a number of known actinonin analogue inhibitors like the

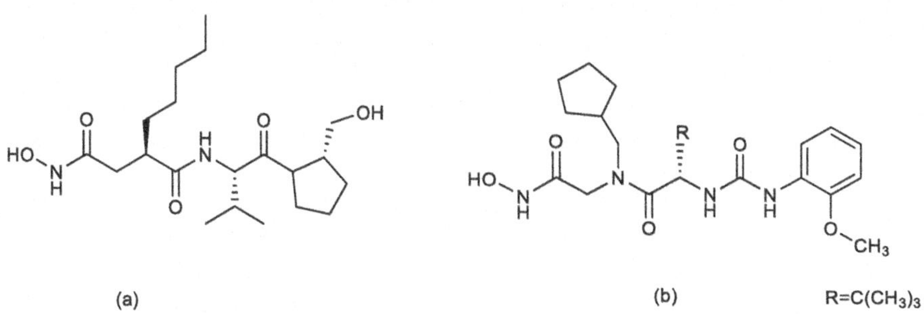

Fig. 4.1 Structure of actinonin (**a**) and an actinonin analogue incorporating a urea group (**b**)

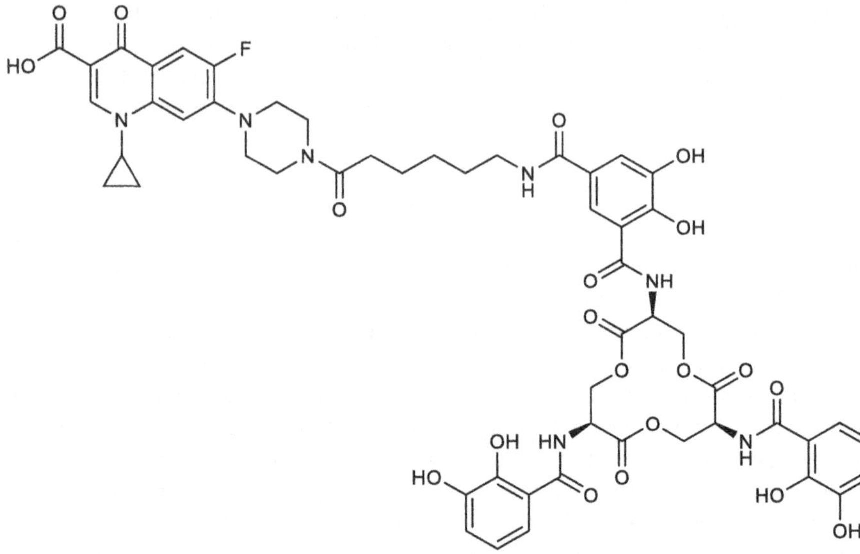

Fig. 4.2 Structure of a enterobactin-ciprofloxacin prodrug

urea-embedded compound shown in Fig. 4.2b, might be a feasible approach to the prodrug.

A variation on the lactamase-initiated ring opening trigger to set off the formation of products by subsequent chemical processes could also involve two pharmacophore-containing leaving groups. Grant and Smyth (2004) have discussed and referred to their approach with a cephalosporin-based dual release system with a view to releasing two bioactive products. In this work β-lactamase induced cleavage of the fused lactam moiety then led to two chemical processes involving elimination in one case to release one of the products from the 3′-position and a ring sulfur-mediated intramolecular displacement at an S-aminosulfenimine in the 7-position to expel the second product. This approach has considerable potential for the selective release of antibacterial products with single or multi-targeting characteristics.

The type (i) cleavage strategy has also been applied to target pathogenic Gram-negative bacteria in some elegant work by the Miller group in the USA (Liu et al. 2018). In this case conjugation of a siderophore-cephalosporin to an oxazolidinone antibacterial via a carbamate linking group was designed and synthesised. The oxazolidinone antibacterial is released after β-lactam cleavage followed by elimination in the enamine produced to ultimately release carbon dioxide and the oxazolidinone. A key aspect is selectivity of targeting through use of the siderophore and active transport across the outer membrane in clinical isolates of the Gram-negative bacterium *Acinetobacter baumannii*. Selective β-lactamase-mediated release of the oxazolidinone antibacterial then occurs followed by diffusion through the inner membrane and intracellular release where it can inhibit protein synthesis through ribosomal binding (Leach et al. 2007). The conjugate (a synthetic sideromycin) was also highly active against strains of *Acinetobacter baumannii* which produce large quantities of the lactamase ADC-1, as well as against *Escherichia coli* and *Pseudomonas aeruginosa.* The siderophore-mediated active transport coupled with β-lactamase induced drug release is a powerful combination for overcoming drug resistant bacteria. Replacement of the oxazolidinone with ester-linked ciprofloxacin, a strategy also used by Evans et al. (2019), while retaining the attached siderophore moiety, is suggested to be of interest as long as the ester linkage remains intact prior to active uptake of the conjugate.

Targeted antibacterial delivery in an inactive enterobactin-ciprofloxacin conjugate prodrug (Fig. 4.2) was achieved through the enterobactin siderophore uptake mechanism in an *Escherichia coli* pathogenic strain which expressed the *iro*A gene cluster. After uptake, hydrolysis of the siderophore enterobactin moiety by a particular cytoplasmic hydrolase IroD in this strain then results in intracellular release of the active antibacterial (Neumann et al. 2018). Such bacterial enzyme specificity could be used as the basis for the development of narrow spectrum antibacterials.

Some antibiotics may also produce reactive oxygen species (ROSs) which could complement other activity pathways, although this ROS proposal has been criticized by others (Owens 2013). Interestingly hydrogen peroxide released by *Streptococcus pneumoniae* inhibits inflammasomes resulting in immune suppression (Erttmann and Gekara 2019). The bacteria avoid attack by the immune system, but how *Streptococcus pneumoniae* maintains high resistance itself to hydrogen peroxide is still incompletely understood.

Reactive oxygen species could be useful, however, in triggering the release of antibacterials in situ. In the anti-cancer context, Noh and co-workers reported on the use of a hydrid drug responsive to dual stimuli which increased the death of cancer cells via enhancement of oxidative stress (Noh et al. 2015). The hybrid was based on a substituted carbonate diester moiety, with one substituent group being the source of 1,4-benzoquinone methide (from hydrogen peroxide oxidation of a boronate substituted benzyl unit) and the second substituent group releasing the ROS generator cinnamaldehyde on exposure to acidic conditions in the cancer cell. The quinone methide would deplete the concentration of the natural oxidant glutathione and overall an increase in oxidative stress results. Conceivably this approach could be adapted to achieve multi-action antibacterial effects, perhaps through enhancement of aPDT and using the sensitivity of aryl boronate esters to hydrogen peroxide and the release of a benzoquinone methide intracellularly. Interestingly in this general context, reactive oxygen species also seem to be involved in the antibacterial activity of two 1,4-benzoquinones isolated from an unusual source, the venom of the scorpion *Diplocentrus melici.* These benzoquinones were found to have good bactericidal activity against *Staphylococcus aureus*, and one of the componds, 6-methoxy-2,3-dithiomethyl-1,4-benzoquinone, also showed potent activity against both drug-sensitive and drug resistant *Mycobacterium tuberculosis* both in vitro and in vivo while apparently not damaging lung epithelia. Initiial results indicated this compound acts through the generation of reactive oxygen species and subsequent attack by these on a range of possible targets. These benzoquinones also interfered with glutathione in a cell based assay (Carcamo-Noriega et al. 2019).

In the cleavable type i design manifold, a range of other structural variants are possible apart from those already discussed if A and B are incorporated in different structures, for example ligand–metal ion complexes where the metal ion might be A and one of the ligands having a recognition element B. Such complexes may then be activated through target-ligand binding. Rutledge and co-workers are working towards an interesting realisation of one aspect of this target binding activation approach with the design concept of target-activated metal complexes as a magic bullet type antibacterial therapy. They are developing a metal ion-cyclam complex to which is attached another ligand with an N-atom also complexed to the metal ion in the inactive state. When this pendant ligand binds to its biological target this reveals the active metal ion in the cyclam or other complex which can then exert other biological activity in situ (Spain et al. 2018). The biological target does not necessarily have to be an enzyme but it could be. Initial studies indicate good anti-TB activity with prototype compounds with a cyclam metal-complexing core and different pendant groups. Such compounds are radically different from other antibacterials and incorporate structural features and components with potential for multi-activity expression. Mode of action studies are being pursued. This work also uses a Zebra fish model for assessment of in vivo activity (*Mycobacterium marinarum*) and there are possibilities for the extension of this model to other bacterial pathogens (Yu et al. 2016; Spain et al. 2018).

If A and B are incorporated in monocyclic or bicyclic type systems then on activation through A---B cleavage could give active single molecules with say A′ and B′ exposed in a linked structure able to express multiple binding properties through the exposed groups. Many possibilities present themselves with these cyclic type i variants. For example one could consider spring loaded or umbrella prodrug designs as generalized in Fig. 4.3 where cleavage of the A-B link by a bacterial enzyme would result in the molecule springing open to reveal the active pharmacophoric groups or moieties A′ and B′. With this compact design, strain release or increased conjugation or bond weakness could be used to drive the unravelling. Kinetics considerations with respect to rates of release would need to be carefully considered though. Multi-activity possibilities could result if the umbrella handle moiety A′ was involved in at least a single interaction and the tip moiety B′ was capable of a dual action binding at two target sites at the same time. Realistion of this type of prodrug should be feasible.

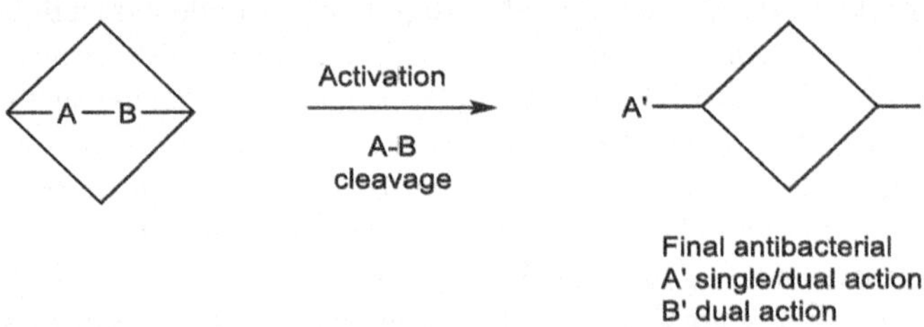

Fig. 4.3 General representation of a spring-loaded prodrug

4.2.4 Cleavable Type II

This cleavable type in its basic form can be represented by:

$$\mathbf{A\text{- -}B\text{- -}C \rightarrow A\text{- -}B' + C'}$$

With this general type of prodrug design, cleavage of one of the linking groups could reveal a dual action hybrid (A--B′) together with another molecule C′ capable of interacting with a third separate target involved with the bacterium. In this case the selective advantage hinges on the active components being released in or near the bacterium. While the initial molecule may have no inherent antibacterial activity, the aim would be to expose active components in high concentration on reaching the bacterium utilizing bacterially specific enzymes for the initial activation and then non-enzymatic steps if required.

One possible molecular realisation of this cleavable type ii prodrug is illustrated by (I) (Scheme 4.1) (Bremner 2017) which incorporates the dual release prodrug design. In this case a monocyclic β-lactam triggering unit, to which C is attached, could be linked to a fluoroquinolone core B also associated with a further target recognition moiety A (Scheme 4.1). If the β-lactam in (I) was cleavable by a β-lactamase enzyme or on interaction with a PBP, then, by analogy with some *N*-sulfonyloxy-β-lactam inhibitor studies of Mourey et al. (1999) and Swarén et al. (1999), subsequent elimination step (1) might then follow to give the fluoroquinolone derivative (II) (a substituted ciprofloxacin with the further recognition unit A attached) in which the key 3-carboxylic acid would be present. The iminium group produced in the remaining acyl enzyme unit (or the ketone hydrolysis product, Mourey et al. 1999) from elimination step (1) might then serve in turn to activate a second elimination step (2) to give C′ (III) if the attached fragment C had an appropriately positioned electron acceptor group. C′ could be designed to interact with another bacterial target site and express another synergistic activity.

While (II) and (III) (Scheme 4.1) could potentially be separately prepared and then used in a two-component combination, the prodrug design strategy suggested would provide a means to help overcome any pharmacokinetic issues when used in vivo to treat resistant bacterial infections. From a synthetic perspective, the *N*-acyloxy substituted β-lactam unit in (I) should be accessible based on the established carbodiimide and Mitsunobu reaction methodology for other mono β-lactams (Swarén et al. 1999). The review on monocyclic β-lactams as antibacterials and β-lactamase inhibitors may also provide a guide to rational design of new triple acting prodrugs taking note of regions which can be altered structurally without affecting activity (Decuyper et al. 2018).

In the type (ii) construct even greater targeting can be achieved if the product C′ is a gaseous compound with actual or possible multi-actions as for example with nitric oxide or carbon monoxide. Kelso, Rineh and colleagues have published some compelling work on the release of nitric oxide from a cephalosporin-diazeniumdiolate prodrug (Scheme 4.2) utilising β-lactamase activation or PBP

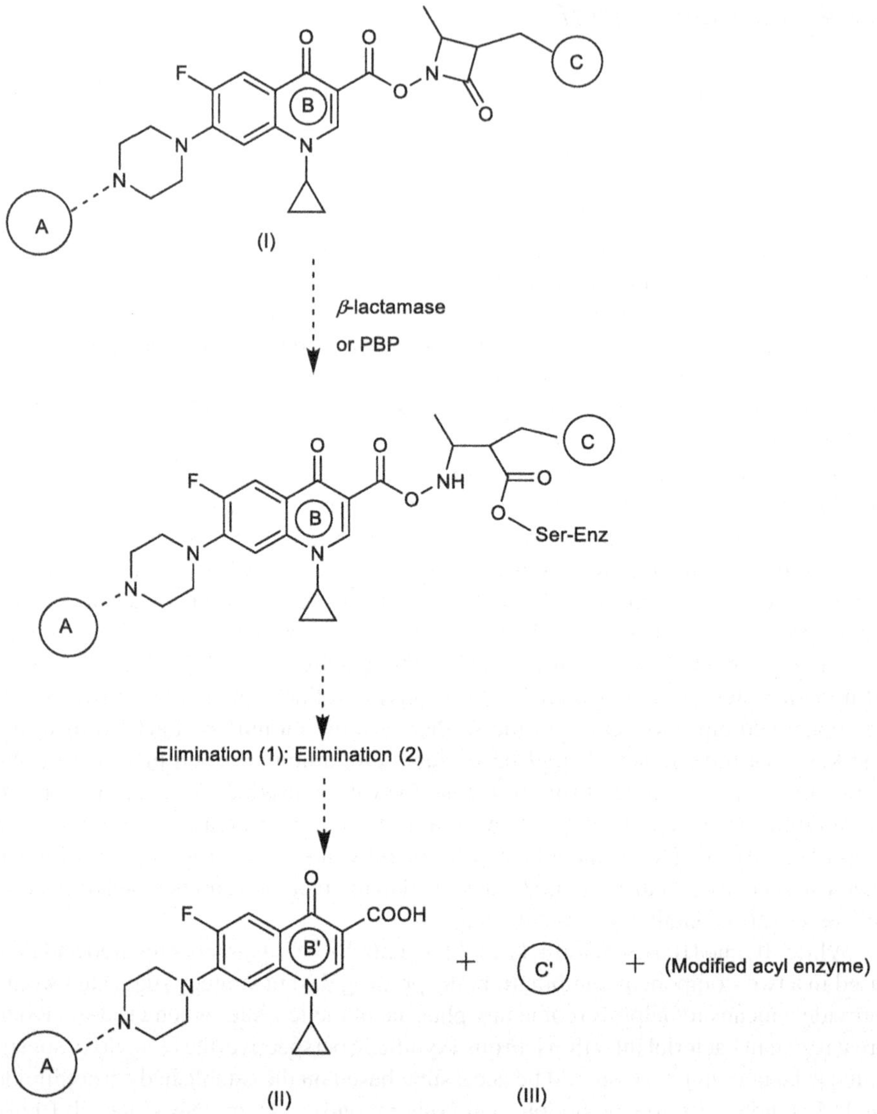

Scheme 4.1 Possible fragmentation pathway of the proposed type ii prodrug (I) after β-lactamase or PBP binding activation

binding (Allan et al. 2017). The first generation work on targeted cephalosporin-3′-diazeniumdiolates as NO-donor prodrugs which can disperse bacterial biofilms was discussed in Barraud et al. (2012) and involved a terminal diethylamino group in the diazeniumdiolate unit.

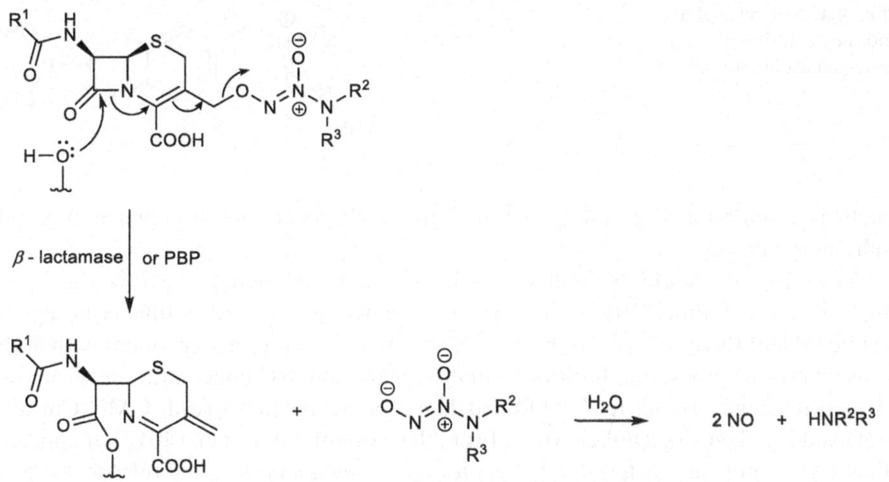

Scheme 4.2 Illustration of the release of nitric oxide from a cephalosporin-diazeniumdiolate prodrug

Further evolution of these NO-releasing prodrugs incorporating a Ceftazidime–based side chain gave compounds which showed good dual antibacterial and anti-biofilm activity against *Pseudomonas aeruginosa,* particularly with the terminal amine component being a 4-(2-aminoethyl)piperidino unit (Rineh et al. 2020). The nitric oxide released after PDP binding was thought to mediate the biofilm disruption while planktonic cells exposed were susceptible to the direct antibacterial properties of the cephalosporin-β-lactam. With other pathogens, variable potency from poor to moderate was displayed. Unfortunately no activity was seen against *Mycobacterium tuberculosis,* but LDT (L,D-transpeptidase) protein binding (Levine and Beatty 2019) together with outer membrane penetration problems may be compromising issues in this case with the NO releasers. Cell wall issues in *Mycobacterium tuberculosis* were also noted by Wivagg et al. (2014) with β-lactams targeting the peptidoglycan structure. It has been suggested that enzymes integral to the modification and metabolism of the *Mycobacterium tuberculosis* (Mtb) peptidoglycan should be looked at more intensively as alternative antibacterial targets (Catalão et al. 2019).

In terms of structural variations on the NO-releasing cephalosporin prodrugs an interesting change might be to replace the carboxylic acid group in the cephalosporin by a boronic acid unit to inhibit the potent Mtb β-lactamase (BlacC) and perhaps improve membrane penetration as long as PBP binding is retained. The synthesis of such compounds is also likely to stimulate the development of some new chemistry which could be capable of wider application. With the boronic acid group, ionization at physiological pH gives a possible equivalent of the essential carboxylate anion substituent in the 6-membered cephalosporin ring system. There are a number of reports of boronic acids as β-lactamase inhibitors (Hevener et al. 2013) with better cellular uptake through ion trapping, one example being the boronic acid-based

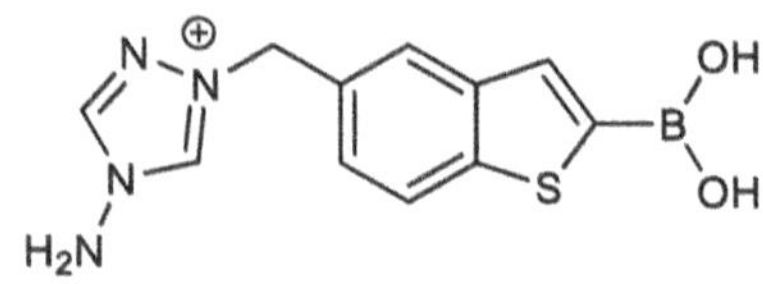

Fig. 4.4 Structure of a boronic acid-based β-lactamase inhibitor

lactamase inhibitor (Fig. 4.4) which had good cell penetration together with good inhibitory activity.

With NO it should be noted that it is said to be unreactive towards most biomolecules (Gilmer 2017) although it is known as a potent antibacterial agent (Hibbard and Reynolds 2019a, b). It's effects are concentration dependent with high concentrations promoting biofilm formation while low NO concentrations promote dispersal when NO binds to H–NOX and then later involvement of c-di-GMP (Cutruzzolà and Frankenberg-Dinkel 2016). From the work of Allan et al. (2017), it appears that NO is not antibacterial for *Streptococcus pneumoniae* after release from a cephalosporin-nitric oxide donor prodrug but it does modulate bacterial signalling as well as influencing metabolic processes which expose *Streptococcus pneumoniae* to antibiotics. Putting the other amine product to work, for example if it were part of a fluoroquinolone with a piperidinyl group, would seem attractive from the point of view of exploiting this susceptibility to antibiotics. The molecular weight of the required prodrug precursor would be problematic though for oral administration. Furthermore, such an antibiotic would be released in the cell wall and would need to diffuse through to inside the cell before accessing the gyrase and topoisomerase sites.

One potential disadvantage with this NO-mediated approach is the formation of toxic *N*-nitrosamines after NO release. However, *N*-nitrosoproline is one nitrosamine which is not toxic (as referenced in Hibbard and Reynolds 2019a) so it could be embedded in the R^3-N-R^2 terminal of the diazenium diolate moiety in the cephalosporin prodrug design (Scheme 4.2). To obviate *N*-nitrosamine formation, one might look to bypass having a terminal amino unit and consider using other substituents at this position which still act as leaving groups. As a thought experiment one might consider incorporating boron-based functionality in a new type of diazenium diolate derivative like $O–N=N^+(O^-)–O–B(OH)_2$ with the terminal oxygen attached to the cephalosporin methylene group by a C–O bond. With this group two moles of nitric oxide could still be released plus borate, which is also known to have further antibacterial actions.

Another potential issue with nitric oxide is that it has a short half-life and it is hard to control it's concentration. It has been suggested that stable nitroxide species as NO mimetics might be a better option. Although nitroxide still needs to be shown to be a mimetic group, nitroxide derivatives can inhibit biofilm growth and promote dispersal (Verderosa et al. 2017). Nitroxides would offer other opportunities for type ii prodrug design and it should be relatively straightforward to integrate a nitroxide like Tempol into the cephalosporin C3-side chain via an ether linkage from which it should be released on lactam ring opening.

In an extension of the work of Hibbard and Reynolds (2019b), utilisation of intracellular nitroreductase enzymes in anaerobic bacteria is indicated as well to release nitric oxide together with an amine-containing antibacterial like ciprofloxacin to intensify the antibacterial effects through further actions. A drawback may be that ciprofloxacin could form the *N*-nitroso derivative with other detrimental effects or with reduced antibacterial activity. For example *N*-nitrosonorfloxacin is less active as an antibacterial compared with norfloxacin (Adjei et al. 2006). This approach may be useful in treating tuberculosis with a judicious choice of the fluoroquinolone (Schluger 2013).

Interesting prodrug hybrids of quorum sensing inhibitors with nitric oxide donors have also been investigated such as fimbrolide-nitric oxide donor hybrids of the nitroxy and diazeniumdiolate types with different linkers or direct attachment. These hybrids also offer further opportunities for potentiating interactions (Kutty et al. 2013). They seem to act through dual pathways mediated through biofilm morphology alteration and increasing susceptibility to NO. Other hybrids of acylated homoserine lactone and nitric oxide donors have also been reported with good antibacterial properties based on the complementary inhibition of quorum sensing and virulence factors (Kutty et al. 2015).

Apart from the release of nitric oxide, another possibility is the release of carbon monoxide (CO). Considerable work has been undertaken to develop prodrugs which could release carbon monoxide after suitable activation and good reviews on the topic are those by Ling et al. (2018) and Ji and Wang (2018). Carbon monoxide and carbon monoxide releasing molecules (CORMs) have shown activity against both Gram-positive and Gram-negative bacteria. Carbon monoxide acts as a bactericide primarily through inhibiting the respiratory chain via stopping adenosine triphosphate supply (Ling et al. 2018). A number of the CORMs are metal ion complexes and their bactericidal activity may also stem from this. Ways to achieve selectivity in the actions of CO-releasing prodrugs are also important considerations. In the anti-cancer context, an interesting way to achieve this involved enrichment-triggered prodrug activation through mitochondria targeted delivery of doxorubicin or CO prodrugs which are biorthogonally activated by a cycloaddition reaction with a substituted cyclooctyne derivative (Zheng et al. 2018). Mitochondrial targeting was achieved through the inclusion of terminal triphenylphosphonium ion functionality on substituent groups on the prodrugs and on the cyclooctyne unit; such functionality is known to greatly increase enrichment in the mitochondria of the molecules to which it is attached. This approach is elegant conceptually but it may be difficult to selectively adapt this design to antibacterial demands (no mitochondria) and have selective concentration only in bacterial cells, however it deserves further consideration in the antibacterial prodrug context. Further interesting possibilities present themselves in the case of visible light-induced CO release from transition metal-free flavonoid-based systems (photo-CORMs) (Anderson et al. 2015). One might, for example, attach a cephalosporin unit via ether formation using the 3-hydroxy group in the flavone which could then be released after interaction of the prodrug with β-lactamase, and in the presence of light result in possibly more bacterially localised co-release of CO. Extension of the light-induced reaction to activation by radiation in the near infrared region would be likely to improve the applicability of the approach to deeper infection sites in the host.

Another possible alternative to NO is nitroxyl (HNO), which is well known in the gas phase. It is a weak acid, with the conjugate base NO^- being a reduced form of NO and isoelectronic with dioxygen. When HNO is released in the aqueous medium of the bacterial cell it would also be the major species present rather than NO^- as the pKa has been assessed as 11.5, rather than 4.7 as thought previously (Switzer et al. 2009). However, the deprotonation reaction of HNO is very slow due to intersystem crossing and the kinetic and thermodynamic barriers are high enough to prevent interconversion between HNO and NO. Both species have different chemical reactivities with NO being a nucleophile favouring electron transfer, while HNO is a good electrophile which favours addition reactions to nucleophiles, particularly thiols with consequences for the biochemical targets attacked (Switzer et al. 2009). Also, if in sufficient concentration locally, HNO dimerises and the dimer (hyponitrous acid) then decomposes to give nitrous oxide and water.

4.2.5 Cleavable Type III

$$\mathbf{A\text{- -}B\text{- -}C} \rightarrow\rightarrow \mathbf{A}' + \mathbf{B}' + \mathbf{C}'$$

In this type iii case sequential cleavage of each linking group would lead ultimately to the release of three separate components A′, B′ and C′. One expression of this type concerns a block copolymer (A) to which the antibiotic gentamicin (B) was attached via a hydrolysable imine linkage to an aromatic aldehyde group in the polymer and then an *N*-diazeniumdiolate (C) attached via the secondary amine group ($-NH-CH_3$) in the gentamicin unit. On exposure to mildly acidic conditions in the bacteria, simultaneous release of nitric oxide (C′) (from diazeniumdiolate hydrolysis) and gentamicin (from imine hydrolysis) resulted in a prolonged antibacterial effect which reduced the viability of a biofilm of *Pseudomonas aeruginosa* by greater than 90% in vitro. There is a synergistic effect from having both gentamicin and nitric oxide present (Nguyen et al. 2016). Nguyen and co-workers have subsequently reviewed nano-nitric oxide delivery systems and their applications, including antibacterial ones (Nguyen et al. 2018).

Another expression of type ii is that seen with a compound (compound 5) in the results presented by Lee et al. (2004). This compound is a tripartite carrier-linked prodrug with two different antibacterial (nalidixic acid and 5-aminoquinoline) moieties at each terminus with a linking TAT peptide unit. Phenylacetic ester groups at each end serve as substrates for bacterial penicillin G amidase (PGA) which on hydrolysis then triggers eliminative release of the two antibacterials plus quinone methides attached to the TAT peptide which then undergo further intramolecular reactions. The compound was very potent against an *Escherichia coli* strain transformed with a PGA gene (not all strains have PGA). One could envisage simpler in-principle versions of this construct with an aryl ether substituting a monobactam at the 4-position and utilizing lactamase- or PBP- induced ring opening triggering the reaction cascade to release one or two antibacterials plus the reactive *para*-quinone methide unit. The quinone methide (either attached to the enzyme or released) may then interact with a number of other bacterial targets as a Michael acceptor and a number of known Gram-positive antibacterials incorporate this functional assembly within their structures, for example the nortriterpene quinone methide celastrol (Moujir et al. 2014) and the novel antibacterial *p*-quinone methide elansolid A3 (Jansen et al. 2011) isolated from the bacterium *Chitinophaga sancti*.

In terms of future design within this general type there are many possibilities. For example, molecule (II) above in Scheme 4.1 (Sect. 4.2.2 cleavable type ii) might include a feature in the linking group to A which was susceptible to cleavage by a later trigger to give A′ and the antibacterial ciprofloxacin as B′. Also a sulfonate ester could be part of the leaving group and release sulfur dioxide as well as another antibacterial with a phenolic group. In addition, $-(CH_2)_nNH_2$ could be incorporated to aid Gram-negative penetration as well as then possibly assisting with release of the active enzyme by intramolecular lactam formation i.e. chemically assisted re-activation.

Another potential variation on this theme which could result in three products with single or multi-targeting abilities might centre on activation of a cephalosporin-based prodrug by a β-lactamase (or PBP) triggering further bond cleavages and the release of *p*-quinone methide (A′), as well as sulfur dioxide (B′) and the hydroxy derivative (C′) (Scheme 4.3). The *p*-quinone methide would also react rapidly with hydroxide ion/water to give *p*-hydroxybenzyl alcohol which could be a bonus as well in view of the known bacteriostatic activity of some phenols for a range of bacteria. Interestingly, endogenous production of the related *p*-cresol has been demonstrated in the bacterial pathogen *Clostridium difficile* together with it's role in affecting gut microbial diversity and membrane integrity in Gram-negative bacteria (Passmore et al. 2018). The release of *p*-quinone methide was also involved in mediating some of the activity in an anti-cancer prodrug (Noh et al. 2015). Other prodrugs can release

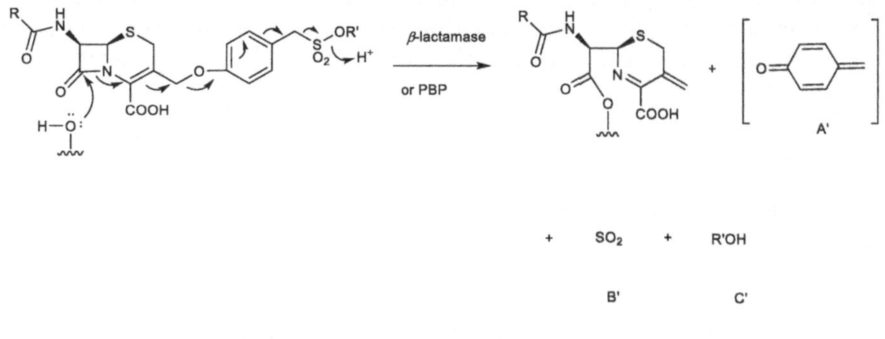

Scheme 4.3 Suggested prodrug precursor and reaction process for the production of three potentially antibacterial products including sulfur dioxide

sulfur dioxide and Wang and Wang (2018) have reviewed recent developments with SO_2 releasing prodrugs with antibacterial activity. Sulfur dioxide and bisulfite have antibacterial properties and like NO, sulfur dioxide is also endogenously produced.

A better alternative to a β-lactamase trigger might be to use a bacterial nitroreductase to initiate the cleavage, for example a *p*-nitrobenzyl goup, as used successfully for nitric oxide release (Hibbard and Reynolds 2019b). Although this restricts use to anaerobic pathogenic bacteria it would bring the molecular weights down significantly. The reduction of nitro groups to amines by nitroreductases can be variable but the factors involved have been helpfully and broadly analysed by Miller et al. (2018).

Instead of using a cephalosporin as the basis for the triggering event, another possibility is to use a simpler monobactam for this purpose. A suggested prodrug construct based on a monobactam is illustrated in Scheme 4.4. After ring opening of the β-lactam, a cascade of non-enzymatic reactions may then proceed as shown to ultimately release the antibacterial ciprofloxacin (A), a sulfonamide (B), and the potentially bioactive vinyl imine-diamino diene (C). The product (B) could be designed to be a sulfonamide antibacterial with the appropriate substituents including one on the sulfonamide nitrogen, while (C) has the potential to act in the vinyl imine tautomeric form as a Michael acceptor for protein-based nucleophilic groups.

One of the keys to the cleavage reactions seen with a number of prodrugs is the generation of a core unit based on an enamine unit of the type: –NH–CH=CH–CH_2–LG where the leaving group contains one or more pharmacophoric units as in a triple acting hybrid. The conjugated imine fragment also released might be designed to target a different bacterial site and express other activity or activities. This a powerful release mechanism which has the potential for further applications in the type iii category. The basic core enamine can also be added to resulting in the generation of moieties like –NH–CH=C(…pharmacophore C)–CH_2–LG (pharmacophore B) by bacterially specific enzymatic means from an enzymatically vulnerable group on N in the prodrug. The pharmacophore B-containing group could be released by subsequent chemical hydrolysis in the cell to afford say an α,β-unsaturated aldehyde unit which might also display multi-reactivity with other nucleophilic sites via Michael-type additions. Alternatively, after release of pharmacophore B, the resultant imine unit may favour tautomerisation to another enamine unit if some stabilisation is provided by an appropriate substituent, followed by release of another leaving group from this second enamine.

4.2.6 Cleavable Type IV

$$\mathbf{A\text{- -}B\text{- -}C\text{- -}X \rightarrow A\text{- -}B\text{- -}C' + X'}$$

With this design type, selective cleavage of one linking group could result in the fully active triple action agent being released near the biological target sites. To add a further action, fragment X′ might also be designed to have a synergistic fourth

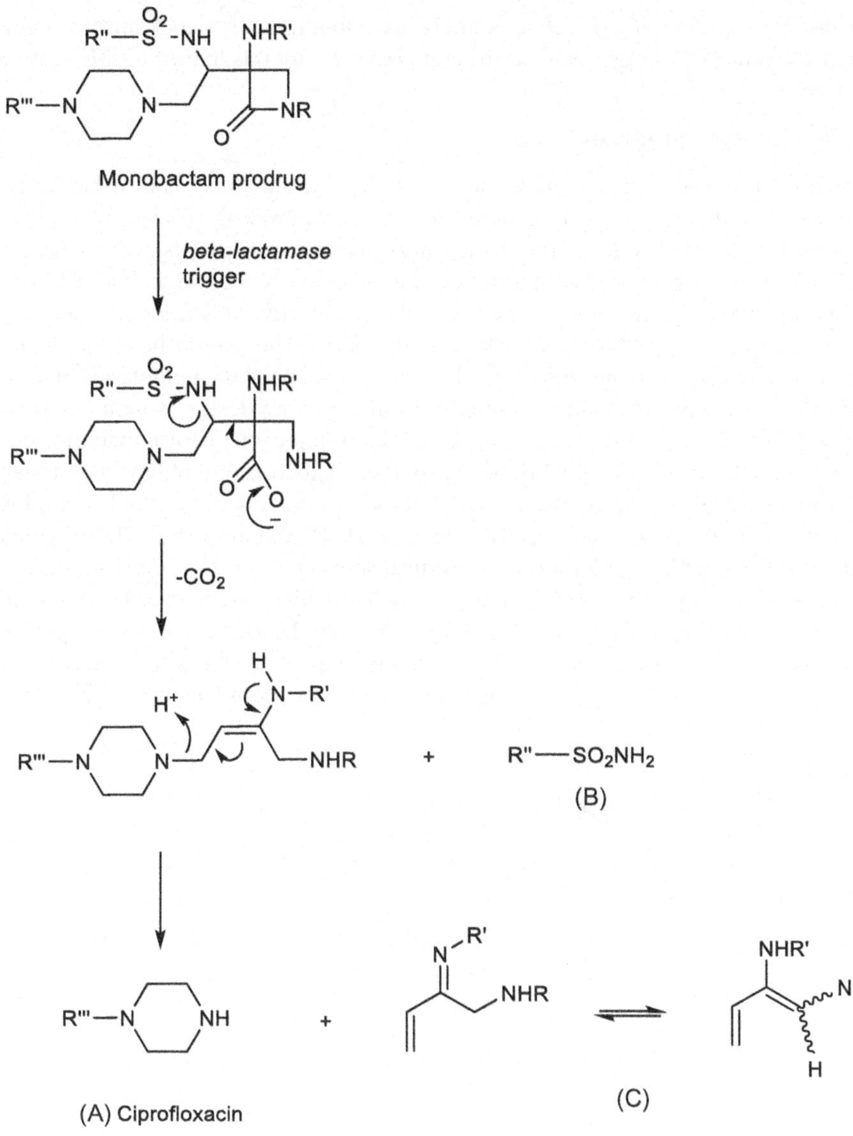

Scheme 4.4 A suggested monobactam-based prodrug and proposed fragmention scheme to release ciprofloxacin and two other potentially bioactive compounds

biological action. The cytoplasmic enzyme peptide deformylase might be considered for cleavage of the terminal C--X linkage (Sangshetti et al. 2015). This enzyme does have some substrate tolerance and the intermediate or final amine generated in the formamide hydrolysis process could then serve as a subsequent chemical trigger to release a triple action antibacterial agent in high concentration intracellularly.

Similarly cleavage of a β-lactam could be used to ultimately expel a triple acting moiety (Scheme 4.5). A generalized structure is shown for this hybrid for illustrative purposes.

Russian doll-type prodrugs

An additional speculative prodrug variation which is recommended for further investigation is one that is not really covered by the above prodrug types although can be viewed as an extension of the Trojan horse-type design discussed in Chap. 3 (Sect. 3.3.3). These proposed compounds are based on what might be called 'Russian Doll-type prodrugs' or molecular assemblies for the selective and sequential delivery of three components at different bacterial target sites. This would be a significant molecular challenge but one which could beneficially expand the boundaries of drug synthesis and design. It would essentially entail a nesting design, which could be unpealed successively to release the drug as such or a precursor for metabolic activation in situ. Some chemical work has been described covering molecular Russian doll motifs based on organic compounds, but these compounds are generally complex molecular assemblies (Cai et al. 2018; Zhang et al. 2019; Gong et al. 2016) which might detract from their exploration in the drug discovery space. However, this design strategy should not be discarded for this purpose particularly with regard to potential multi-action antibacterials. One can envisage adapting the idea to systems requiring for example two bacterial enzymes to sequentially unpeel layer 1 with the release of drug 1, then layer 2 with the release of drug 2. The active components, which might

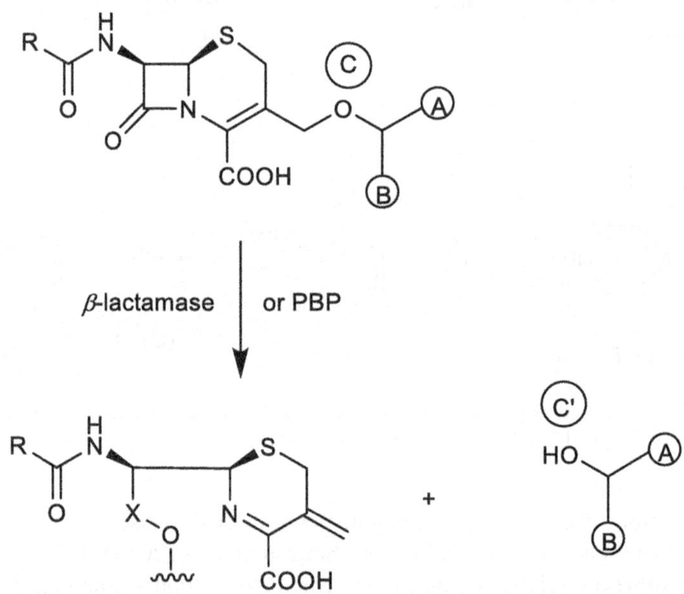

Scheme 4.5 Potential release of a single trihybrid from a cephalosporin prodrug

have one or more activities, would then be released in a potentially time controlled manner which could improve potency and final antibacterial outcomes. It is of note that nature can encapsulate enzymes in nanometric compartments to increase and modulate the activity of the enzyme (Zhang et al. 2019).

4.3 Release Mechanisms and Prodrug Design

Prodrug release mechanisms, which are crucial considerations in prodrug design, can involve enzymatic or non-enzymatic (chemical or physical) processes or combinations of these. In the following sub-sections these mechanisms are discussed in the three main categories: biological, chemical and physical processes.

4.3.1 Biological

A number of bacterial enzymes have been used for prodrug activation and drug release at intracellular, cell wall, and extracellular bioactivation sites with the last being close to or contiguous with the bacterium. Most commonly, bacterially specific β-lactamases (Ehmann et al. 2013) or PBP-penicillin binding proteins have been utilised for the bioactivation process. A range of others have been studied from the bioactivation viewpoint including peptide deformylase, an Fe^{2+} dependent metallo-hydrolase which mediates hydrolysis of a terminal *N*-formyl group from peptides (Sangshetti et al. 2015; Yuan et al. 2001), azoreductase, bacterial esterases, and bacterial amidases (for example penicillin G amidase (PGA). Prodrugs need to be substrates for the bacterial enzymes and preferably not interacting irreversibly at the enzyme active site so that activation can be continued with release of the active products. While enzymes only occurring in bacteria are preferential targets for the activation process, differences between bacterial and human enzymes with a similar function, for example the recently identified human mitochondrial peptide deformylase, can possibly still be exploited to achieve bacterial specificity. Other issues which need to be considered in the prodrug design and activation are the differences in enzymes between Gram-positive and Gram-negative bacteria.

A further consideration is the potential problem of the site of release of the active compound (s) relative to the antibacterial target sites as in the case of release intracellularly while the target(s) may be in the cell wall or vice versa. Anatomical sites can also perhaps be used to advantage for specific release. For example, azo group reducing bacteria are present in the colon (Abet et al. 2017) and it is suggested this enzymatic reducing ability could potentially be utilised for local release and subsequent absorption of a suitable antibacterial or for localised selective treatment of a bacterially-mediated colon infection. The ability of the azo group reducing bacteria to release an active anti-inflammatory drug (5-aminosalicylic acid, which exerts its effect locally) has been realised with the prodrug balsalazide (Fig. 4.5) (Abet et al.

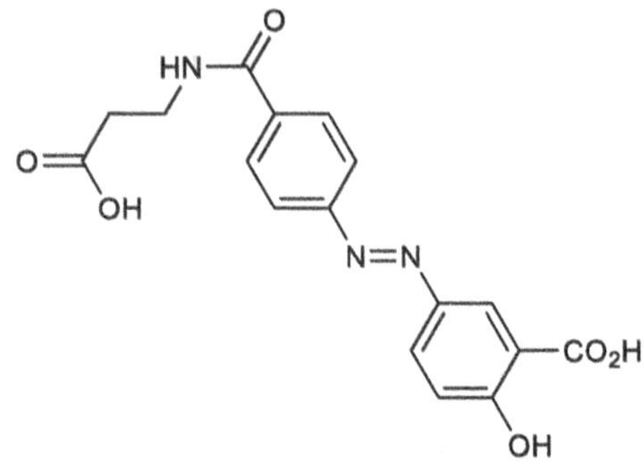

Fig. 4.5 Structure of the azo derivative balsalazide

2017). Balsalazide is used for the treatment of inflammatory bowel disease.

A considerable amount of work has been done on developing inhibitors of peptide deformylase (PDF) as antibacterial agents and, significantly, they cover a variety of different structural types, including both small peptide derivatives as well as non-petidic compounds (Sangshetti et al. 2015; Jain et al. 2005; Lee et al. 2010; Chen and Yuan 2005). There is also some potential for use of this enzyme in prodrug activation. One example includes the activation of 5′-peptidyl derivatives of 5-fluorodeoxyuridine (FdU) by PDF removing the *N*-terminal formyl unit from the dipeptidic group to then trigger release of FdU (a thymidylate synthetase inhibitor) via intramolecular displacement utilising the freed terminal primary amino group as a nucleophile. Unfortunately, though, only weak (*Escherichia coli*) to moderate (*Staphylococcus aureus*) antibacterial activity was evident with this type of prodrug possibly due to poor transport across the cell membrane(s) (Wei and Pei 2000); FdU is a potent antibacterial agent against Gram-positive pathogens like MRSA and vancomycin-resistant *Enterococci*, as well as having good activity in vitro against *Escherichia coli* (Oe et al. 2020).

In view of the task the enzyme is required to do, a range of peptides or proteins must be tolerated as substrates for PDF as long as they have the terminal formyl-Met-Ala-Ser (fMAS) or formyl-Met-Ala (fMA) moieties present. However, formyl-Met-Leu-p-nitroanilide can act as a substrate (Nguyen and Pei 2008), binding in the conserved and important S'_1 pocket. It would thus seem that a range of other substrates for the enzyme to deformylate should be possible to develop. This would then release a free amino group to initiate further chemically-induced release of active antibacterials. One example might be formyl–NH–C(CH_2–CH_2–SMe)=CH–CH_2–N(H)–cephalosporin or fluoroquinolone (at the 7-amino position on the β-lactam in the cephalosporin or at the piperazinyl terminal N in a fluoroquinolone like ciprofloxacin) or formyl–NH–CH(CH_2–CH_2–SMe)–CO–CH_2–N(H)–antibacterial where keto-enol tautomerism might afford the intermediate enamine for the elimination to proceed. Another variation on this might be to include a further antibacterial prodrug moiety on the cephalosporin which could be released on subsequent β-lactamase or PBP cleavage of the β-lactam, thus resulting in triple or more activity. A further study of the structural requirements for PDF substrates rather than inhibitor structural determinants could be beneficial in this area since PDF is required to induce a reaction in the prodrug rather than inhibit it.

There may also be a possibility for the released compound from deformylation to act as an inhibitor of the next enzyme, methionine amino peptidase, which is required to remove the methionine unit. This enzyme is an essential one in bacterial and human cells so the compound would need to have selective inhibition properties, but this might be possible based on it's targeted release within the bacterial cell. A review of advances in this area of bacterial methionine peptidase inhibitors has been published by Helgren et al. (2016), and from enzyme structural studies significant differences have been identified between the bacterial and human enzymes (Helgren et al. 2016).

Another approach to consider is that of using bacterial cell peptidase to activate a prodrug after assisted transport into the cell. This approach was used successfully with eneamide prodrugs of inhibitors of the essential bacterial enzyme 1-deoxy-D-xylulose-5-phosphate (DXP) synthase. This enzyme is essential in bacterial central metabolism and is found in most pathogenic bacteria but it is not present in animals. Work is being undertaken into enamide prodrugs of the selective mechanism-based acylphosphonate inhibitors of this enzyme as potent antibacterials (Bartee et al. 2019). The prodrug design was based on the natural product dehydrophos and the peptidic enamide prodrugs were transported into the cell via peptide transporter, OppA. Inside the cell the peptidase cleaved the peptide units to give an enamine which could then be hydrolysed to the active acetyl phosphonate DXP inhibitors (Bartee et al. 2019). This approach has considerable potential for improved intracellular prodrug delivery to bacterial cells via the OppA transporter. Different peptidic prodrugs could be transported including ones with an additional antibacterial leaving group appropriately attached to an enamide moiety for the spatio-temporal release of a dual acting hybrid for synergy with the acetyl phosphonate DXP inhibitor. In this context a design extension to replace the phosphonate moiety by a boronate unit could be pursued with a view to potentially increasing DXP inhibitory potency or invoking a second activity.

Targeting of anaerobic over aerobic bacteria could be achieved with involvement of nitroreductase in the former for prodrug activation and release of the active component(s). If the actives are only released once inside the bacterium where the bacterial enzyme resides this will result in selectivity of action, assuming the prodrug is not affected by external host enzymes or pH and doesn't have other interactions with non-bacterial targets resulting in unwanted side effects. In this context consideration might be given to designing a piperazine *N*-substituted nitro-vinyl ciprofloxacin derivative which could be activated by a bacterial nitroreductase to release ciprofloxacin plus ultimately an α,β-unsaturated aldehyde. Such an aldehyde might be expected to covalently interact with a number of intracellular protein sites or other sites, in addition to the gyrase/topoisomerase targeting ciprofloxacin, providing an expression of multiple activity from the prodrug (Scheme 4.6). Further development of this conceptual approach might include nitro-azole analogues.

Pretomanid (Fig. 4.6), an orally bioavailable, nitroimidazo-oxazine based prodrug, is activated by a deazaflavin (cofactor F_{420})-dependent nitroreductase. The

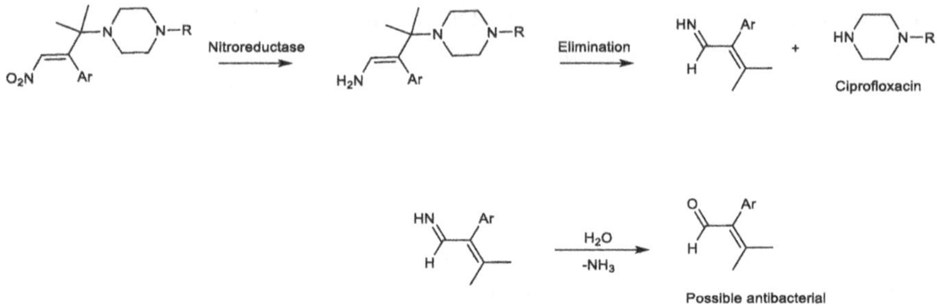

Scheme 4.6 Suggested reaction scheme for the release of antibacterial components by nitroreductase

Fig. 4.6 Structure of the prodrug pretomanid

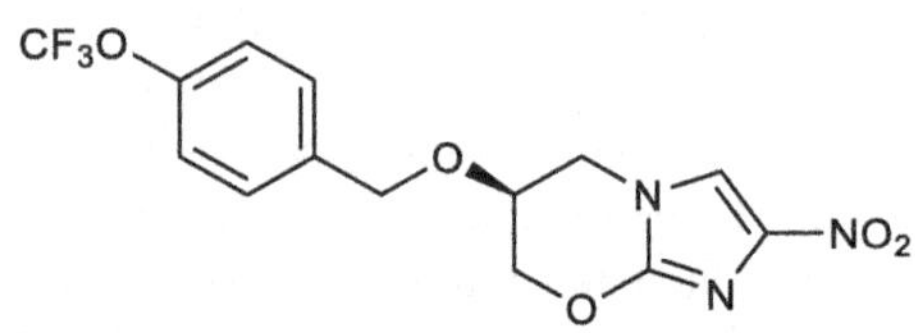

drug released inhibits cell wall mycolic acid biosynthesis and also furnishes nitric oxide which can poison the respiratory chain. While the NO mechanism of action here is not fully resolved it is likely to be multidimensional. Details on the proposed NO producing mechanism are given in the paper by Thompson et al. (2017) which deals with anti-TB bicyclic nitroimidazoles. Pretomanid, in combination with bedaquiline and linezolid, was approved by the FDA in August 2019 for the treatment of a specific type of pulmonary tuberculosis which is highly resistant to treatment.

The use of a β-lactamase as a trigger for prodrug activation and drug release continues to be widely investigated. Work recently reported by Evans et al. (2019) (see also Sect. 4.2.2 in this chapter) covers refinement of the β-lactamase induced release of ciprofloxacin from an ester linked conjugate. The difference from previous compounds of this type is that the prodrug is designed to have no or little inherent antibacterial activity and would thus not release the ciprofloxacin until exposed to lactamase producing pathogens. Activity similar to ciprofloxacin itself was seen with an *Escherichia coli* isolate expressing β-lactamases. Such selectively activated prodrugs have potential clinical advantages in terms of less broad brush disruption to microbiota and reducing selection for resistance pressures. This prodrug approach may also help to obviate the severely detrimental toxic effects of ciprofloxacin (and other fluoroquinolones) seen in a small percentage of the population systemically exposed to fluoroquinolone antibacterials in the treatment of bacterial infections (Marchant 2018).

Non-β-lactams working somewhat like a β-lactam have been developed and these, like avibactam, react reversibly with some serine β-lactamases (Wang et al. 2016).

The process involves ring opening via the attack of the Ser_{67} residue in the lactamase (for example OXA-10) on the cyclic urea carbonyl with cleavage of the C–N(OSO_2OH) bond in the urea moiety, followed by intramolecular *N*-acylation to reform avibactam and the free serine residue. Such mechanistic knowledge could be used to design in structural features to allow release of another antibacterial after the enzyme-mediated ring opening, perhaps via an intramolecular nucleophilic displacement process. However such designs do not appear to have been reported in the literature as yet.

Also of interest from a design point of view is the development of a prodrug for avibactam as an oral antibiotic, previously only administered intravenously. Having an oral form is a significant advance (Li 2018). Two of these avibactam prodrugs are shown in Fig. 4.7. In the prodrug form, avibactam is released in vivo from its stable *O*-neopentyl-derived sulfate ester promoiety after esterase-mediated hydrolysis of the terminal ester group (for example the benzyl or ethyl ester) and subsequent intramolecular displacement of avibactam with the gem-dimethyl group aiding this internal displacement by bringing the carboxylate nucleophile closer to the sidechain methylene group site. The dimethyl group was also important for stabilising the prodrug by sterically hindering intermolecular nucleophilic attack on this methylene group (Gordon et al. 2018).

Another quite different type of biological trigger which can be mentioned here involves the controlled release of an antibacterial from a pre-formed liposome. Pornpattananangkul et al. (2011) have shown that vancomycin can be released from liposomes stabilised by chitosan-modified gold nanoparticles when they are exposed to bacterial toxins. When near *Staphylococcus aureus* the secreted toxins produced by the bacterium then insert in the liposome forming pores which allows the vancomycin to be released in sufficient concentration locally to inhibit bacterial growth. This type of process could have wider potential for specific antibacterial release, including multi-targeting hybrids, particularly for the treatment of resistant topical infections, and perhaps other bacterial infections if the somewhat complex structural features can be simplified or streamlined.

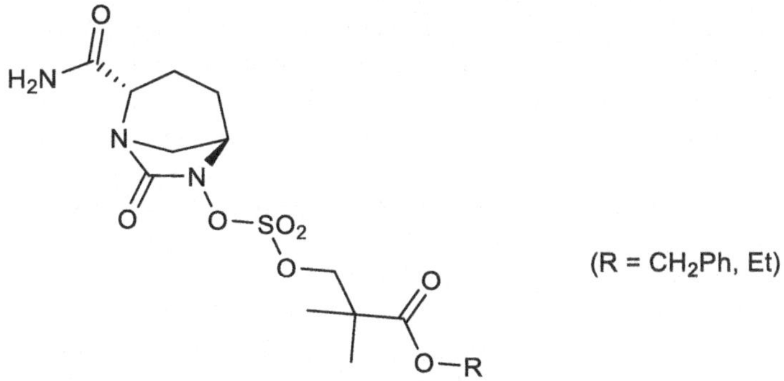

Fig. 4.7 Structures of two *O*-neopentyl-derived sulfate ester prodrugs of avibactam

4.3.2 Chemical

Bio-orthogonal prodrugs

The term bio-orthogonal chemistry is used to differentiate any chemical reaction which takes place in living systems without interfering with inherent biochemical processes. Such reactions have been applied to activate prodrugs through strain promoted bio-orthogonal chemical processes and subsequent drug release. This has been developed for the release of a number of drugs including cytotoxic agents typified by the insightful work of Matikonda, Gamble and co-workers (Matikonda et al. 2015, 2018). Their prodrug activation strategy involved five steps with a key initial one being a strain-promoted 1,3-dipolar cycloaddition reaction of an aryl azide (linked in turn to the inactive cytotoxin unit) to a substituted *trans*-cyclooctene to give an unstable 1,2,3-triazoline derivative which could fragment and rearrange with nitrogen loss to give an imine linked prodrug unit. Acid catalysed hydrolysis of the imine then triggered a further 1,6-elimination to release the active cytotoxic drug. This type of activation approach is adaptable in principle to antibacterial release as shown by Czuban et al. (2018) who have reported on targeted treatment of local bacterial infections using the catch and release strategy. Their strategy requires an initial local injection of material with a tetrazine attached to an alginate gel at the target infected site in vivo, followed by a systemic dose of the prodrug, then a concentration phase when the the two components come in contact and 'catch', and finally prodrug activation. The 'catch' in this case involves an inverse electron demand Diels–Alder reaction between the strained *trans*-cyclooctene amide-linked antibiotic prodrug unit and the 1,2,4,5-tetrazine diene component with release of nitrogen. The activation and release stage is mediated by spontaneous isomerization (tautomerisation) in the dihydrodiazine ring which then promotes elimination of carbon dioxide and release of the antibiotic (daptomycin or vancomycin) near the infection site. This then leaves an aromatic 1,2-diazine unit (after tautomerisation) fused to the cyclooctane and attached to the gel. The number of stages in this approach is a little problematic and the challenge will be to simplify and make the process more compact, but at least the principles have been established.

There are other bio-orthogonal approaches to prodrug cleavage (Weiss et al. 2015) involving non-biological, physical or chemical initiators to ultimately release the active component or components from the prodrug which could also be useful if the prodrug can be concentrated in bacteria or at their surface. A good review on biorthogonal approaches to 'on demand' prodrug activation is that by Ji et al. (2019). The emphasis here is on the click and release approach, but one needs to watch overly complicated designs and a reliance on intermolecular reactions which can be problematic in terms of getting reactants together in sufficient concentration in vivo. Secondary intramolecular reactions would thus be preferable and ways to incorporate this possibility in the overall prodrug design should be further investigated.

Combinations of biological and chemical triggers can also be effective and these usually involve enzymatic cleavage of a prodrug followed by further chemical reaction of the intermediate produced to give other bioactive products. Scope exists also for possibilities with the reverse triggers, that is with the active drug or drugs being released in the second stage by enzymatic action after an initial chemical or photochemical reaction activation step.

An example of biologically-induced release followed by chemical release has been noted for the broad spectrum anti-fungal agent Isavuconazole (Rautio et al. 2018). In this case the quaternary salt prodrug Isavucazonium sulfate undergoes initial activation by esterase enzyme-mediated hydrolysis of a remote ester unit. The alcohol unit revealed then proceeds spontaneously to a rapid intramolecular cyclisation reaction which triggers in turn a *N*-dealkylation step to release the active Isavuconazole. Application of this strategy to prodrug design in antibacterial settings should also be assessed.

4.3.3 Physical

Physically-initiated release or activation mechanisms are also important in the prodrug setting. Extensive work has been done on photo-induced bio-orthogonal chemistry for spatio-temporal control in delivering molecules for interaction with biological targets (Li et al. 2020) as part of the photopharmacology field (Arrue and Ratjen 2017; Velema et al. (2014). Applications of phototriggered targeting in nanomedicine (Arrue and Ratjen 2017) remain to be fully explored in the antibacterial area as are applications of photo-switchable antibiotics in interfering with quorum sensing (Feringa 2017). This suggests in turn that photoswitching could be combined with drug release from a prodrug although this would mean visible- or near IR-light induced bond cleavage not UV light induced reactions in view of the negative effects of UV light. However, initial design ideas might usefully come from an assessment of known UV-photocleavable protecting groups and then modification of the structural features to include chromophores which would absorb light mainly in the visible or preferably the NIR region.

For photo-activated prodrugs, new designs could come from compounds known to undergo photoisomerisation rather than light induced cleavage reactions producing other products. For example, one might base new designs on analogues of a substituted dihydropyrene system. Such a dihydropyrene system has been described with a donor $N(CH_3)_2$ and acceptor (NO_2) substituent which are coupled in the coloured form and on one photon near infrared excitation is isomerised to a colourless cross-conjugated product in which these groups are uncoupled. This isomer then converts back thermally (T) to the coloured form (a negative T-type photochromic system) (Klaue et al. 2018). From these results it could be of interest to look at heterocyclic analogues, for example the new diaza analogue proposed in Scheme 4.7a. In this conjugated system the donor tertiary amino and the nitro acceptor group are coupled. A 6π electrocyclic internal ring opening to give the cross-conjugated system where the donor and acceptor groups are uncoupled (Scheme 4.7b) on irradiation with near infrared light would be expected to occur. This could then expose the resultant α,β-unsaturated nitroalkene moiety to nucleophilic attack by a number of sulfur- or

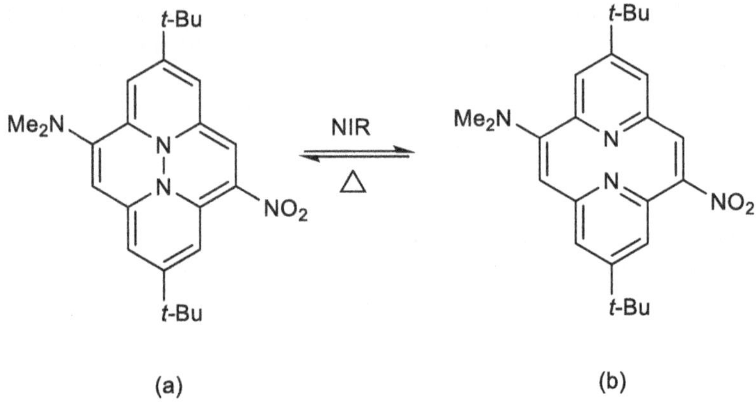

Scheme 4.7 Proposed new diaza analogue of a dihydropyrene system (**a**) and its cross conjugated photoisomer (**b**) from near infrared activation

nitrogen-based nucleophiles on bacterial proteins, a process which could compete with the thermal reversal step. To mitigate the clash of the pyridine lone pairs it is likely a twist in the macrocyclic ring system would be imposed further activating the β-position to such nucleophilic addition.

Light is a good externally applied spatio-temperal control element although there are limitations with respect to viable wavelengths (600–1200 nm) due to absorption and optical scattering controlling intensity; light of wavelength 600 nm can penetrate about 1 cm of tissue while at 1200 nm it is about 2 cm. Optic fibres can allow light access to most body organs potentially offering light-induced reactions for organ-located bacterial infections where no other options exist.

Light-mediated localised heating (gold particle-based) can also be used to trigger the release of drugs based on the work of the Kohane group and disclosed in Zhan et al. (2016). They describe the controlled release of tetrodotoxin from a gold nanorod-liposome entity on brief exposure to tissue penetrant near IR light. This is a good demonstration of light-induced drug delivery on demand mediated by localised heating of the gold nanorods on the liposome surface then thermally disrupting the liposome. In this construct the liposome itself can be viewed as a type of prodrug. This methodology should be adaptable to antibacterial hybrid release or related prodrug release where the liposome was accumulated.

Gold nanoclusters have also now been described for bacterial detection as well as actions against bacteria importantly involving the production of reactive oxygen species which can attack different targets (Tang et al. 2020). In a review on the topic of nano-strategies to combat multidrug resistant bacteria, a range of mechanisms of action by nanoparticles against bacteria are noted, as well as other potential issues with regard to therapeutic treatment. Toxicity to mammalian cells, much of which is not clear, may be a drawback (Baptista et al. 2018).

The effects of ultrasound waves (vibration frequencies higher than 20 kHz) on liposomes and nanometal particles is also being actively investigated in the context of the development of antibacterial treatments. Ultrasonic wave absorption can cause an increase in temperature in tissue from local mechanical compression and decompression events which can disrupt cell membranes. Complicated non-thermal effects of ultrasound can also occur which can result in mechanical stresses and increases in temperature (Canavese et al. 2018). Developments in this area include the intensification or collection of low intensity ultrasound by a suitable molecule or nanoparticulate system concentrated selectively in bacteria and much of this work has focussed on the use of gold nanoparticles. Nanoparticle-assisted ultrasound in cancer therapy is also under active study (Canavese et al. 2018) and the results from this area have ramifications for multi-active prodrug antibacterial design and ways to avoid host cell damage while attacking pathogenic bacteria in vivo.

The effects of ultrasound directly on bacteria is also relevant as perhaps one can use these effects to improve antibacterial outcomes (Monsen et al. 2009). An inhibitory effect on Gram-negative bacteria, particularly on *Escherichia coli*, has been observed, but Gram-positive bacteria were more resistant (Monsen et al. 2009). However, in other work (Liao et al. 2018) both E*scherichia coli* and *Staphylococcus aureus* were irreversibly affected, perhaps because of the different experimental conditions. Also it has been shown that low-intensity and low-frequency ultrasound combined with the antibiotic tobramycin on multidrug resistant *Escherichia coli* biofilms shows promise in significantly decreasing bacterial viability of biofilms with the ultrasound disruption aiding antibiotic penetration (Hou et al. 2019).

Other interesting possibilities present themselves in prodrug design if one 'action' is induced by a physical phenomenon. And in this context the bacterial mechanosensitive channels are of great interest. These channels, amongst other functions, are vital for mitigating the effects of hypo-osmotic shock in bacteria (Booth 2014).

Antibacterial compounds which interact with the bacterially specific mechanosensitive ion channel of large conductance (MscL) are known like the novel drug Ramizol (Fig. 4.8) (Iscla et al. 2015; Rao et al. 2016; Wolfe et al. 2018). Ramizol is active against methicillin-resistant *Staphylococcus aureus* (Iscla et al. 2015) and against

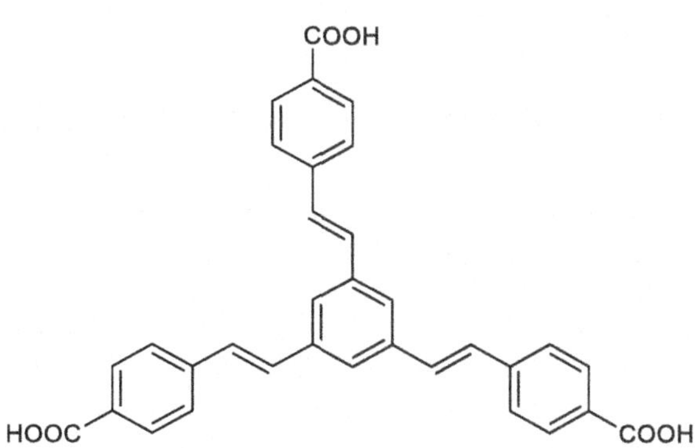

Fig. 4.8 Structure of the antibacterial ramizol which interacts with the mechanosensitive ion channel (MscL)

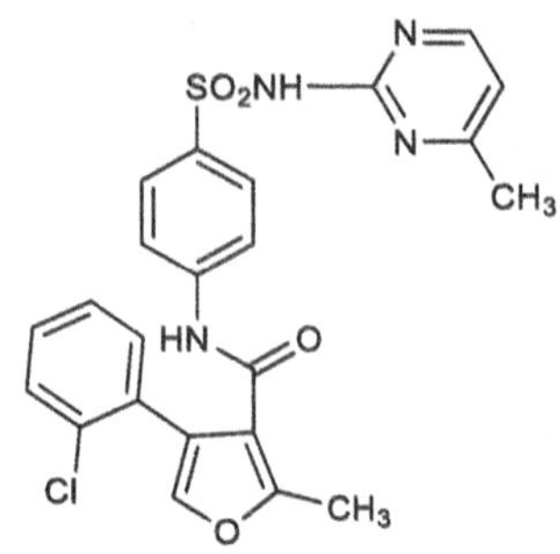

Fig. 4.9 Structure of the MscL interacter compound 011

Clostridium difficile, with potential for development as an efficacious treatment of *Clostridium difficile* colitis (Rao et al. 2016). Ramizol slows the growth of bacteria through lowering the threshold at which the MscL channel opens and also lengthening the channel opening times (Rao et al. 2016). There is considerable evidence that in *Escherichia coli* membrane tension in the lipid bilayer directly activates the mechanosensitive channel MscL to open but details on how this is mediated are not fully understood (Iscla and Blount 2012; Haswell et al. 2011). An X-ray structure on the MscL homolog from *Mycobacterium tuberculosis* has indicated an oligomerised structure for this channel (Chang et al. 1998; Yoshimura et al. 1999). Potential exists in this area to incorporate Ramizol or analogues within a gold particle-based liposome for targeting bacteria, and then apply ultrasound-induced liposome disruption for release of the drug. The drug might then interfere with the functioning of mechanosensitive channels and the ultrasound could possibly induce pore formation (Babakhanian et al. 2018) resulting in possible increased antibacterial efficacy.

Other work (Wray et al. 2019) has concluded that compounds which interact specifically with MscL and increase gating have antibacterial activity thus further confirming MscL as a feasible antibacterial target. The sulfonamide compound 011 (Fig. 4.9) mediated it's antibacterial activity in *Escherichia coli* through increasing MscL activity and not through the folate pathway as seen for other sulfonamides (Wray et al. 2019). It also suggests new possibilities for multiply active hybrids and prodrugs.

4.4 Metabolism Activated Multi-targeting

Metabolism activated multi-targeting (MAMUT) has potential for new antibacterial design. The approach is based on an active drug and its metabolite(s) having interactions with different biological targets in a synergistic manner. The strategy was proposed initially by Mátyus and Chai (2016) and they demonstrated a proof of concept with a compound SZV-1287 which inhibited the enzyme semicarbazide-sensitive amine oxidase and was also converted in vivo to a known cyclooxygenase inhibitor, oxaprozin. SZV-1287 and oxaprozin display synergistic activity. Although a prodrug is not specifically involved initially they are presumably involved as intermediates on the steps through to oxaprozin. MAMUT is related through ultimate outcomes to the bioprecursor prodrug classification but is precluded from this classification since, formally, a prodrug is not active until activated. In nature, metabolic activation to give an active multi-targeting antibacterial from a precursor can also occur as exemplified by the lantibiotic lacticin 481. This antibiotic, which is produced by some strains of the Gram-positive bacterium *Lactococcus lactis*, arises from a ribosomally-produced prepeptide which is then post-translationally modified by the enzyme lantibiotic synthetase (LctM) in two stages and finally released after proteolytic cleavage of a leader peptide unit (Xie et al. 2004).

In applying the MAMUT approach in the multi-action antibacterial context one would need to look for a suitably functionalized antibacterial molecule, perhaps a natural product (Ho et al. 2018), with a metabolically sensitive functional group or moiety whose likely metabolites might potentiate the action of the starting drug by interacting with another target or targets. If the metabolic transformation was unique to the bacterium this would be a further plus. This could be quite difficult to achieve in molecular terms but not impossible. One could consider for example a starting compound such as the *N*-formyl structure (Scheme 4.8a), which could have multi-activity. If the starting material was also bioactive then it falls under MAMUT. Metabolism of the terminal *N*-formamide moiety by bacterial peptide deformylase could conceivably generate a primary amino group which might then internally displace a group C′ with one activity and a cyclic hybrid product (Scheme 4.8b) with a pendant A-B group capable of interacting with two other target sites either on the one macromolecule or at one site each on two different macromolecules. Product C′ might be a monobactam or contain a pharmacophoric unit which could inhibit another aspect of bacterial protein synthesis. Apart from designing for synergistic interactions, considerations would need to include the rate of hydrolysis of *N*-formamides independent of peptide deformylase. Such hydrolysis would need to be very slow.

4.5 Conclusion

Both carrier-linked and bioprecursor prodrugs have been developed for release of compounds with multi-targeting capability in the antibacterial area. Various designs and triggering mechanisms have been assessed but there is much more scope for new prodrugs with better selectivity and different release strategies. Particular emphasis on releasing multi-active hybrid compounds as well as multi-active gaseous molecules should be prioritised in future research.

The ultimate goal would be to design and develop orally bioavailable, targeted Gram-negative and Gram-positive penetrant prodrugs as precursors for multi-targeting agents active against both resistant and non-resistant strains and with a low propensity for resistance development. This is a major challenge which might best be met by focussing on separate sub-categories of compounds with perhaps

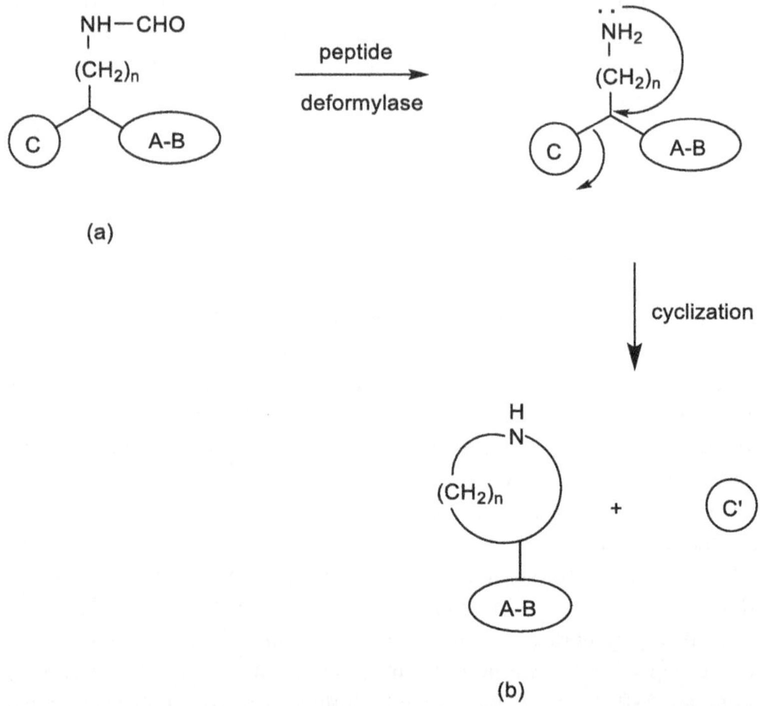

Scheme 4.8 Proposal for application of metabolism activated multitargeting (MAMUT) to an *N*-formamide (**a**) and producing a dual targeting antibacterial hybrid (**b**) plus a further compound with a different target

more circumscribed antibacterial profiles in terms of the pathogens targeted. A key design consideration will continue to be making maximum use of each atom or atom combinations in the molecules by looking to have at least dual capabilities for each group or at releasing more than 1 mol of the active agent from 1 mol of the precursor as with nitric oxide release from diazeneniumdiolates. Less should indeed be more.

Prodrugs are the final arm in the antibacterial multi-action design processes described in this book. The basic intersectionality of these design components are highlighted in Fig. 4.10. These components include known or potential drug combinations with two or more different molecules (red box), and alternative lead compounds from natural or other sources (green box), all informing the design of non-cleavable single molecule hybrids with multi-actions or other multi-action molecules (blue box). Prodrug designs (yellow/orange box) then hinge on releasing these last compound types from single molecules in a selective manner. It should also be noted that only the basic interactions are shown in Fig. 4.10 for clarity and further interactions can also be useful for example between leads and combinations or leads and prodrug design input.

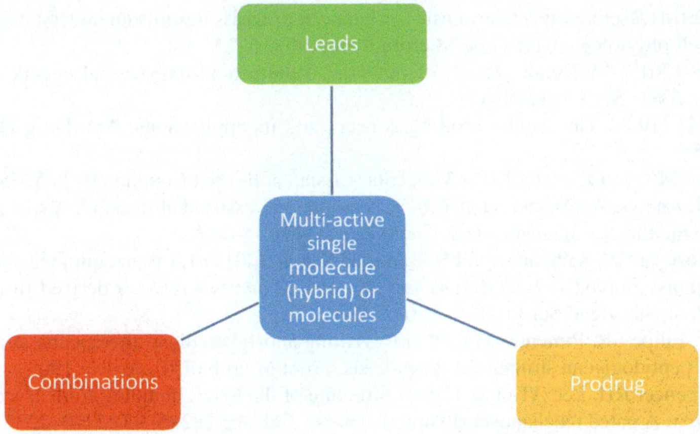

Fig. 4.10 Intersectional diagram for multi-active antibacterial design

There are many possibilities for the design and development of radically new, triple- or higher action antibacterials from prodrugs and, while the medicinal chemical and other disciplinary challenges are demanding, such antibacterial agents could ultimately help to meet the antibiotic resistance threat and improve health care outcomes in the future.

References

Abet V, Filace F, Recio J et al. (2017) Prodrug approach: An overview of recent cases. Eur J Med Chem 127:810–827

Adjei MD, Heinze TM, Deck J et al. (2006) Transformation of the antibacterial agent norfloxacin by environmental mycobacteria. Appl Environ Microbiol 72 (9):5790–5793

Allan RN, Kelso MJ, Rineh A et al. (2017) Cephalosporin-NO-donor prodrug PYRRO-C3D shows β-lactam-mediated activity against *Streptococcus pneumoniae* biofilms. Nitric Oxide 65:43–49

Anderson SN, Richards JM, Esquer HJ et al. (2015) A structurally-tunable 3-hydroxyflavone motif for visible light-induced carbon monoxide-releasing molecules. ChemistryOpen 4:590–594

Arrue L, Ratjen L (2017) Internal targeting and external control: phototriggered targeting in nanomedicine. ChemMedChem 12 (23):1908–1916

Babakhanian M, Yang L, Nowroozi B et al. (2018) Effects of low intensity focused ultrasound on liposomes containing channel proteins. Sci Rep 8 (1):17250–17250

Baptista PV, McCusker MP, Carvalho A et al. (2018) Nano-strategies to fight multidrug resistant bacteria-"A Battle of the Titans". Front Microbiol 9:1441–1441

Barraud N, Kardak BG, Yepuri NR et al. (2012) Cephalosporin-3'-diazeniumdiolates: targeted NO-donor prodrugs for dispersing bacterial biofilms. Angew Chem Int Ed 51 (36):9057–9060

Bartee D, Sanders S, Phillips PD et al. (2019) Enamide prodrugs of acetyl phosphonate deoxy-d-xylulose-5-phosphate synthase inhibitors as potent antibacterial agents. ACS infectious diseases 5 (3):406–417

Bio M, Nkepang G, You Y (2012) Click and photo-unclick chemistry of aminoacrylate for visible light-triggered drug release. Chem Commun 48 (52):6517–6519

Booth IR (2014) Bacterial mechanosensitive channels: progress towards an understanding of their roles in cell physiology. Curr Opin Microbiol 18 (100):16–22

Bremner JB (2017) Molecular design of potential triple-action antibacterial agents and related pro-drugs. SWU Sci J 33 (1):1–19

Bundgaard H (1989) The double prodrug concept and its applications. Adv Drug Deliv Rev 3 (1):39–65

Cai K, Lipke MC, Liu Z et al. (2018) Molecular russian dolls. Nat Commun 9 (1):5275

Canavese G, Ancona A, Racca L et al. (2018) Nanoparticle-assisted ultrasound: A special focus on sonodynamic therapy against cancer. Chem Eng J 340:155–172

Carcamo-Noriega EN, Sathyamoorthi S, Banerjee S et al. (2019) 1,4-Benzoquinone antimicrobial agents against *Staphylococcus aureus* and *Mycobacterium tuberculosis* derived from scorpion venom. Proc Natl Acad Sci 116 (26):12642–12647

Catalão MJ, Filipe SR, Pimentel M (2019) Revisiting anti-tuberculosis therapeutic strategies that target the peptidoglycan structure and synthesis. Front Microbiol 10: Article 190

Chang G, Spencer RH, Lee AT et al. (1998) Structure of the MscL homolog from *Mycobacterium tuberculosis*: A gated mechanosensitive ion channel. Science 282 (5397):2220–2226

Chen D, Yuan Z (2005) Therapeutic potential of peptide deformylase inhibitors. Expert Opin Investig Drugs 14 (9):1107–1116

Cutruzzolà F, Frankenberg-Dinkel N (2016) Origin and Impact of Nitric Oxide in *Pseudomonas aeruginosa* Biofilms. J Bacteriol 198 (1):55–65

Czuban M, Srinivasan S, Yee NA et al. (2018) Bio-orthogonal chemistry and reloadable biomaterial enable local activation of antibiotic prodrugs and enhance treatments against *Staphylococcus aureus* infections. ACS Cent Sci 4 (12):1624–1632

Das N, Dhanawat M, Dash B et al. (2010) Codrug: An efficient approach for drug optimization. Eur J Pharm Sci 41 (5):571–588

Decuyper L, Jukič M, Sosič I et al. (2018) Antibacterial and β-lactamase inhibitory activity of monocyclic β-lactams. Med Res Rev 38 (2):426–503

Domalaon R, Idowu T, Zhanel GG et al. (2018) Antibiotic hybrids: the next generation of agents and adjuvants against Gram-negative pathogens? Clin Microbiol Rev 31 (2):e00077–00017

Ehmann DE, Jahic H, Ross PL et al. (2013) Kinetics of avibactam inhibition against Class A, C, and D β-lactamases. J Biol Chem 288 (39):27960–27971

Erttmann SF, Gekara NO (2019) Hydrogen peroxide release by bacteria suppresses inflammasome-dependent innate immunity. Nat Commun 10 (1):3493

Evans LE, Krishna A, Ma Y et al. (2019) Exploitation of antibiotic resistance as a novel drug target: Development of a β-lactamase-activated antibacterial prodrug. J Med Chem 62 (9):4411–4425

Feringa BL (2017) The art of building small: From molecular switches to motors (Nobel Lecture). Angew Chem Int Ed 56 (37):11060–11078

Fieulaine S, Rodolphe ADS, Maigre L et al. (2016) A unique peptide deformylase platform to rationally design and challenge novel active compounds. Sci Rep 6:35429

Gilmer JF (2017). Towards gaseous mediator hybrid drugs. In: Decker M (ed) Design of hybrid molecules for drug development. 1st edn. Elsevier, Amsterdam, pp 47–81

Gong W, Yang X, Zavalij PY et al. (2016) From packed "Sandwich" to "Russian Doll": Assembly by charge-transfer interactions in cucurbit[10]uril. Chem Eur J 22 (49):17612–17618

Gordon EM, Duncton MAJ, Gallop MA (2018) Orally absorbed derivatives of the β-lactamase inhibitor avibactam. Design of novel prodrugs of sulfate containing drugs. J Med Chem 61 (22):10340–10344

Grant JW, Smyth TP (2004) Toward the development of a cephalosporin-based dual-release prodrug for use in ADEPT. J Org Chem 69 (23):7965–7970

Haswell ES, Phillips R, Rees DC (2011) Mechanosensitive channels: what can they do and how do they do it? Structure 19 (10):1356–1369

Helgren TR, Wangtrakuldee P, Staker BL et al. (2016) Advances in bacterial methionine aminopeptidase inhibition. Curr Top Med Chem 16 (4):397–414

Hevener KE, Cao S, Zhu T et al. (2013) Special challenges to the rational design of antibacterial agents. In: Desai MC (ed-in-chief) Annual reports in medicinal chemistry, vol 48. Academic Press, San Diego, pp 283–298

Hibbard HAJ, Reynolds MM (2019a) Fluorescent nitric oxide donor for the detection and killing of *Pseudomonas aeruginosa*. J Mater Chem B 7(12):2009–2018

Hibbard HAJ, Reynolds MM (2019b) Synthesis of novel nitroreductase enzyme-activated nitric oxide prodrugs to site-specifically kill bacteria. Bioorg Chem 93:103318

Ho TT, Tran QT, Chai CL (2018) The polypharmacology of natural products. Future Med Chem 10 (11):1361–1368

Hou Y, Yang M, Jiang H et al. (2019) Effects of low-intensity and low-frequency ultrasound combined with tobramycin on biofilms of extended-spectrum beta-lactamases (ESBLs) *Escherichia coli*. FEMS Microbiol Lett 366 (3): fnz026

Iscla I, Blount P (2012) Sensing and responding to membrane tension: The bacterial MscL channel as a model system. Biophys J 103 (2):169–174

Iscla I, Wray R, Blount P et al. (2015) A new antibiotic with potent activity targets MscL. J Antibiot 68 (7):453–462

Jain R, Chen D, White RJ et al. (2005) Bacterial peptide deformylase inhibitors: a new class of antibacterial agents. Curr Med Chem 12 (14):1607–1621

Jansen R, Gerth K, Steinmetz H et al. (2011) Elansolid A3, a unique *p*-quinone methide antibiotic from *Chitinophaga sancti*. Chem Eur J 17 (28):7739–7744

Ji X, Pan Z, Yu B et al. (2019) Click and release: bioorthogonal approaches to "on-demand" activation of prodrugs. Chem Soc Rev 48 (4):1077–1094

Ji X, Wang B (2018) Strategies toward organic carbon monoxide prodrugs. Acc Chem Res 51 (6):1377–1385

Jubeh B, Breijyeh Z, Karaman R (2020) Antibacterial prodrugs to overcome bacterial resistance. Molecules 25 (7):1543

Klaue K, Garmshausen Y, Hecht S (2018) Taking photochromism beyond visible: direct one-photon NIR photoswitches operating in the biological window. Angew Chem Int Ed 57 (5):1414–1417

Kutty SK, Barraud N, Ho KKK et al. (2015) Hybrids of acylated homoserine lactone and nitric oxide donors as inhibitors of quorum sensing and virulence factors in *Pseudomonas aeruginosa*. Org Biomol Chem 13 (38):9850–9861

Kutty SK, Barraud N, Pham A et al. (2013) Design, synthesis, and evaluation of fimbrolide–nitric oxide donor hybrids as antimicrobial agents. J Med Chem 56 (23):9517–9529

Leach KL, Swaney SM, Colca JR et al. (2007) The site of action of oxazolidinone antibiotics in living bacteria and in human mitochondria. Mol Cell 26 (3):393–402

Lee M-R, Baek K-H, Jin HJ et al. (2004) Targeted enzyme-responsive drug carriers: Studies on the delivery of a combination of drugs. Angew Chem Int Ed 43 (13):1675–1678

Lee SK, Choi KH, Lee SJ et al. (2010) Peptide deformylase inhibitors with retro-amide scaffold: Synthesis and structure–activity relationships. Bioorg Med Chem Lett 20 (15):4317–4319

Levine SR, Beatty KE (2019) Investigating β-lactam drug targets in *Mycobacterium tuberculosis* using chemical probes. bioRxiv:2019.2012.2019.881631

Li G (2018) First oral antibiotic booster for four decades could tackle drug resistance. Chem World 15 (12):32

Li J, Kong H, Zhu C et al. (2020) Photo-controllable bioorthogonal chemistry for spatiotemporal control of bio-targets in living systems. Chem Sci 11 (13):3390–3396

Li Y-H, Li Y, Yang P et al. (2010) Design, synthesis, and cholesterol-lowering efficacy for prodrugs of berberrubine. Bioorg Med Chem 18 (17):6422–6428

Liao X, Li J, Suo Y et al. (2018) Multiple action sites of ultrasound on *Escherichia coli* and *Staphylococcus aureus*. Food Sci Hum Well 7 (1):102–109

Ling K, Men F, Wang W-C et al. (2018) Carbon monoxide and its controlled release: Therapeutic application, detection, and development of carbon monoxide releasing molecules (CORMs). J Med Chem 61 (7):2611–2635

Liu R, Miller PA, Vakulenko SB et al. (2018) A synthetic dual drug Sideromycin induces Gram-negative bacteria to commit suicide with a Gram-positive antibiotic. J Med Chem 61 (9):3845–3854

Marchant J (2018) When antibiotics turn toxic. Nature 555 (7697):431–433

Matikonda SS, Fairhall JM, Fiedler F et al. (2018) Mechanistic evaluation of bioorthogonal decaging with *trans*-cyclooctene: The effect of fluorine substituents on aryl azide reactivity and decaging from the 1,2,3-triazoline. Bioconjug Chem 29 (2):324–334

Matikonda SS, Orsi DL, Staudacher V et al. (2015) Bioorthogonal prodrug activation driven by a strain-promoted 1,3-dipolar cycloaddition. Chem Sci 6 (2):1212–1218

Mátyus P, Chai CLL (2016) Metabolism-activated multitargeting (MAMUT): An innovative multitargeting approach to drug design and development. ChemMedChem 11 (12):1197–1198

Miller A-F, Park JT, Ferguson KL et al. (2018) Informing efforts to develop nitroreductase for amine production. Molecules 23 (2):211

Monsen T, Lövgren E, Widerström M et al. (2009) In vitro effect of ultrasound on bacteria and suggested protocol for sonication and diagnosis of prosthetic infections. J Clin Microbiol 47 (8):2496–2501

Moujir L, Montaño NP, Bazzocchi IL (2014) Anti-staphylococcal activity of celastrol. Planta Med 80 (16):P2B17

Mourey L, Kotra LP, Bellettini J et al. (1999) Inhibition of the broad spectrum nonmetallocarbapenamase of class A (NMC-A) beta-lactamase from *Enterobacter cloacae* by monocyclic beta-lactams. J Biol Chem 274 (36):25260–25265

Nani RR, Gorka AP, Nagaya T et al. (2015) Near-IR light-mediated cleavage of antibody-drug conjugates using cyanine photocages. Angew Chem Int Ed 54 (46):13635–13638

Neumann W, Sassone-Corsi M, Raffatellu M et al. (2018) Esterase-catalyzed siderophore hydrolysis activates an enterobactin–ciprofloxacin conjugate and confers targeted antibacterial activity. J Am Chem Soc 140 (15):5193–5201

Nguyen KT, Pei D (2008) High-throughput screening of peptide deformylase inhibitors. Methods Mol Med 142:117–130

Nguyen T-K, Selvanayagam R, Ho KKK et al. (2016) Co-delivery of nitric oxide and antibiotic using polymeric nanoparticles. Chem Sci 7:1016–1027

Nguyen KT, Wu Z, Huang T et al. (2018) Molecular design and medicinal applications of nano-nitric oxide delivery systems. Curr Med Chem 25 (12):1420–1432

Noh J, Kwon B, Han E et al. (2015) Amplification of oxidative stress by a dual stimuli-responsive hybrid drug enhances cancer cell death. Nat Commun 6 (1):6907

Oe C, Hayashi H, Hirata K et al. (2020) Pyrimidine analogues as a new class of Gram-positive antibiotics, mainly targeting thymineless-death related proteins. ACS Infect Dis 6:1490–1500

Owens B (2013) Bacteria-killing dispute casts doubt on antibiotic development. Nat Med 19 (8):954

Parajuli RR, Pokhrel P, Lamichane S et al. (2015) Prodrug as a novel approach of drug delivery- a review. J Drug Deliv Ther 5 (3):5–9

Passmore IJ, Letertre MPM, Preston MD et al. (2018) *Para*-cresol production by *Clostridium difficile* affects microbial diversity and membrane integrity of Gram-negative bacteria. PLoS Pathog 14 (9):e1007191

Pornpattananangkul D, Zhang L, Olson S et al. (2011) Bacterial toxin-triggered drug release from gold nanoparticle-stabilized liposomes for the treatment of bacterial infection. J Am Chem Soc 133 (11):4132–4139

Rajendran L, Knölker HJ, Simons K (2010) Subcellular targeting strategies for drug design and delivery. Nat Rev Drug Discov 9 (1):29–42

Rao S, Prestidge CA, Miesel L et al. (2016) Preclinical development of Ramizol, an antibiotic belonging to a new class, for the treatment of *Clostridium difficile* colitis. J Antibiot 69 (12):879–884

Rautio J, Meanwell NA, Di L et al. (2018) The expanding role of prodrugs in contemporary drug design and development. Nat Rev Drug Discov 17 (8):559–587

Rineh A, Soren O, McEwan T et al. (2020) Discovery of cephalosporin-3'-diazeniumdiolates that show dual antibacterial and antibiofilm effects against *Pseudomonas aeruginosa* clinical cystic fibrosis isolates and efficacy in a murine respiratory infection model. ACS Infect Dis 6:1460–1479

Sangshetti JN, Khan FAK, Shinde DB (2015) Peptide deformylase: a new target in antibacterial, antimalarial and anticancer drug discovery. Curr Med Chem 22 (2):214–236

Schluger NW (2013) Fluoroquinolones in the treatment of tuberculosis: which is best? Am J Respir Crit Care Med 188 (7):768–769

Singh GS (2004) Beta-lactams in the new millennium. Part-II: cephems, oxacephems, penams and sulbactam. Mini Rev Med Chem 4 (1):93–109

Spain M, Wong JKH, Nagalingam G et al. (2018) Antitubercular bis-substituted cyclam derivatives: Structure–activity relationships and in vivo studies. J Med Chem 61 (8):3595–3608

Srinivasarao M, Galliford CV, Low PS (2015) Principles in the design of ligand-targeted cancer therapeutics and imaging agents. Nat Rev Drug Discov 14 (3):203–219

Subbaiah MAM, Meanwell NA (2019) Design and tactical applications of prodrugs to enable effective drug delivery. In: Bronson JJ (ed-in-chief) 2019 Medicinal chemistry reviews, vol 54. ACS Division of Medicinal Chemistry, Washington, USA, pp 333–355

Swarén P, Massova I, Bellettini JR et al. (1999) Elucidation of mechanism of inhibition and X-ray structure of the TEM-1 β-Lactamase from *Escherichia coli* inhibited by a N-sulfonyloxy-β-lactam. J Am Chem Soc 121 (23):5353–5359

Switzer CH, Flores-Santana W, Mancardi D et al. (2009) The emergence of nitroxyl (HNO) as a pharmacological agent. Biochim Biophys Acta Bioenerg 1787 (7):835–840

Tang M, Zhang J, Yang C et al. (2020) Gold nanoclusters for bacterial detection and infection therapy. Front Chem 8:181

Thapa P, Li M, Bio M et al. (2016) Far-red light-activatable prodrug of paclitaxel for the combined effects of photodynamic therapy and site-specific paclitaxel chemotherapy. J Med Chem 59 (7):3204–3214

Thompson AM, Bonnet M, Lee HH et al. (2017) Antitubercular nitroimidazoles revisited: Synthesis and activity of the authentic 3-nitro isomer of pretomanid. ACS Med Chem Lett 8 (12):1275–1280

Velema WA, Szymanski W, Feringa BL (2014) Photopharmacology: Beyond proof of principle. J Am Chem Soc 136:2178–2191

Verderosa AD, de la Fuente-Núñez C, Mansour SC et al. (2017) Ciprofloxacin-nitroxide hybrids with potential for biofilm control. Eur J Med Chem 138:590–601

Wang DY, Abboud MI, Markoulides MS et al. (2016) The road to avibactam: the first clinically useful non-β-lactam working somewhat like a β-lactam. Future Med Chem 8 (10):1063–1084

Wang W, Wang B (2018) SO_2 donors and prodrugs, and their possible applications: A review. Front Chem 6: Article 559

Wivagg C N, Bhattacharyya R P, Hung D T (2014). Mechanisms of β-lactam killing and resistance in the context of *Mycobacterium tuberculosis*. J Antibiot 67, 645–654

Wei Y, Pei D (2000) Activation of antibacterial prodrugs by peptide deformylase. Bioorg Med Chem Lett 10 (10):1073–1076

Weiss JT, Carragher NO, Unciti-Broceta A (2015) Palladium-mediated dealkylation of *N*-propargyl-floxuridine as a bioorthogonal oxygen-independent prodrug strategy. Sci Rep 5:9329

Wermuth CG, Ganellin CR, Lindberg P et al. (1998) Glossary of terms used in medicinal chemistry. Pure Appl Chem 70 (5):1129–1143

Wolfe C, Pagano P, Pillar CM et al. (2018) Comparison of the in vitro antibacterial activity of Ramizol, fidaxomicin, vancomycin, and metronidazole against 100 clinical isolates of *Clostridium difficile* by broth microdilution. Diagn Microbiol Infect Dis 92 (3):250–252

Wray R, Iscla I, Kovacs Z et al. (2019) Novel compounds that specifically bind and modulate MscL: Insights into channel gating mechanisms. FASEB J 33 (3):3180–3189

Xie L, Miller LM, Chatterjee C et al. (2004) Lacticin 481: In vitro reconstitution of lantibiotic synthetase activity. Science 303 (5658):679–681

Yoshimura K, Batiza A, Schroeder M et al. (1999) Hydrophilicity of a single residue within MscL correlates with increased channel mechanosensitivity. Biophys J 77 (4):1960–1972

Yu M, Nagalingam G, Ellis S et al. (2016) Nontoxic metal–cyclam complexes, a new class of compounds with potency against drug-resistant *Mycobacterium tuberculosis*. J Med Chem 59 (12):5917–5921

Yuan Z, Trias J, White RJ (2001) Deformylase as a novel antibacterial target. Drug Discov Today 6 (18):954–961

Zawilska JB, Wojcieszak J, Olejniczak AB (2013) Prodrugs: a challenge for the drug development. Pharmacol Rep 65 (1):1–14

Zhan C, Wang W, McAlvin JB et al. (2016) Phototriggered local anesthesia. Nano Lett 16 (1):177–181

Zhang D, Ronson TK, Greenfield JL et al. (2019) Enantiopure [Cs+/Xe⊂Cryptophane]⊂FeII4L4 hierarchical superstructures. J Am Chem Soc 141 (20):8339–8345

Zheng Y, Ji X, Yu B et al. (2018) Enrichment-triggered prodrug activation demonstrated through mitochondria-targeted delivery of doxorubicin and carbon monoxide. Nat Chem 10 (7):787–794

Chapter 5
Future Possibilities

"Think outside the square, but swim between the flags."
—John Bremner.
On Australian surf beaches, red and yellow horizontally-banded flags are used to designate safe swimming areas but you are still very free to think expansively.

Abstract There are many possibilities for the future design of new multi-targeted antibacterials based on developing knowledge of synergistic combinations and new modes of action being discovered. This final chapter includes a comparative assessment of the combination, hybrid and prodrug approaches, together with a discussion of the potential for using new drug combinations, especially incorporating new modes of action, in new hybrid or prodrug designs. A concluding section covers some more general thoughts on the need to think well outside the square in the design and development of antibacterials to help in the fight against the ever increasing threat of resistance.

J. Bremner, *Multiple Action-Based Design Approaches to Antibacterials*,
https://doi.org/10.1007/978-981-16-0999-2_5

5.1 Introduction

This is a more speculative chapter aimed at building on ideas and suggested research directions from the previous chapters. It is not possible to cover all aspects of possible future directions but particular growth areas have been chosen which indicate, or point to, new possibilities in the longer term for small molecule designs potentially with multiple site interactions.

5.2 New Combinations and Single Molecules with Multi-activity Development Potential

5.2.1 Combinations

A range of alternative possibilities are foreseen in this area investigating new combinations of agents not previously assessed. Much scope also exists for the inclusion of dual or triple action hybrids in the mix as well as multi-action prodrugs which may release two antibacterial agents on activation. Dual action single molecules may also be non-hybridic in design where only one recognition feature is known to interact with a bacterial target, and the other feature is not known to interact with a bacterial target but from a chemical viewpoint could possibly do so i.e., possibly active features. Similar considerations could be applied in the prodrug sphere.

For higher order interactions from dual combinations the development and incorporation of new triple action hybrids is indicated. Other agents in the mix could focus more on indirect complementary actions like efflux pump modulation. Such modulation might be realised through using other known or newly designed efflux pump inhibitors. With these studies it will be important to engage a range of structural analysis techniques, including cryo-electron microscopy (cryo-EM), to inform and guide the design of pump inhibitors as well as proposed hybrids incorporating structural features for interacting with the pumps together with exploring the effects of hybridisation on binding to other bacterial targets. Cryo-EM is likely to be a particularly powerful technique for such studies especially with the advance to atomic resolution (Herzik 2020). Cryo-EM was integral to determining the structure of a multidrug efflux pump from the very problematic bacterial pathogen *Acinetobacter baumannii* to a resolution of 2.98 Å (Su et al. 2019). The use of cryo-EM in drug discovery more generally, including successes as well as limitations and future directions, has been covered in an incisive review by Renaud et al. (2018). This review includes a reference to work describing the cryo-EM structure of the *Escherichia coli* 70S ribosome in complex with the elongation factor Tu to which the antibiotic kirromycin is also bound (Fischer et al. 2015). Kirromycin is a potent antibiotic which exerts its protein synthesis inhibition though binding to the Tu factor.

The cryo-EM technique also helped to uncover the mechanism of action of the macrocyclic antibiotic fidaxomicin against *Mycobacterium tuberculosis* (Boyaci et al. 2018). It was found that fidaxomicin interferes with RNA polymerase in this mycobacterium by acting like a wedge keeping the enzyme in an open state thus preventing the essential movement required to lock the promoter DNA in the required active site. While fidaxomicin is active against *Mycobacterium tuberculosis* in vitro this is not the case in vivo after oral administration as it is not absorbed from the intestine to the blood stream. However, from these molecular mode of action details, appropriate structural changes may be possible to facilitate oral absorption of the drug while maintaining the RNA polymerase interference.

The development of new drug combinations for application in synergistic antibacterial disease therapies is also likely to be aided by cryo-EM as a result of the detailed molecular understanding of drug-target interactions for different drugs it can potentially provide (Scapin et al. 2018).

5.2.2 Hybrid Molecule Possibibilities

There are many opportunities for further antibacterial hybrid development. One of these opportunities would be around building on small molecule inhibitors targeting tRNA-(N^1G37)methyltransferase (TrmD) and other enzymes in the bacterial epitranscriptome. One such compound, identified through a compound library search, was a pyridine-pyrazole-piperidine compound (Fig. 5.1) with an amide linkage to an indole unit. This compound was shown to be binding at both the SAM (S-adenosyl-L-methionine-binding pocket and the tRNA binding area in *Pa*TrmD in the epitranscriptome resulting in potent inhibition of this key *Pseudomonas aeruginosa* enzyme (Zhong et al. 2019). Thus a dual mode of action is involved which then raises the question whether this compound could be derivatised to include a pharmacophoric unit capable of a third mode of binding to this enzyme (resistance development might then be very difficult) or other such enzymes in the epitranscriptome thus shutting down essential biochemical processes. Co-ordinating spatial and temporal issues would then not be as difficult perhaps. The interesting collection of different heterocyclic units in this inhibitor offers a number of opportunities for the selective introduction of a suitable functional group for triple site binding or to improve the cellular antibacterial activity characteristics. The new structural scaffold seen in this compound (Fig. 5.1) provides further reinforcement of the emphasis by Walsh and Wencewicz in their prospective review article a number of years ago on the need to find new antibiotic scaffolds (Walsh and Wencewicz 2014).

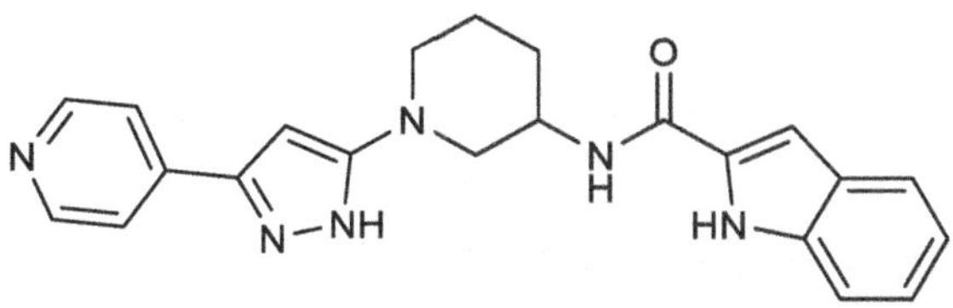

Fig. 5.1 A dual binding action based inhibitor of TrmD

Lipid formation/function processes are potentially other good target areas for new hybridic compounds. For example compounds targeting three of these processes with one including the enzyme LpxC in Gram-negatives would be of great interest (Liu and Ma 2018). There has been quite a lot of work done on LpxC inhibitors, although there seems to be no human clinical studies of such compounds as yet. The highly conserved zinc-dependent enzyme UDP-(3-*O*-(*R*-3-hydroxymyristoyl))-*N*-acetylglucosamine deacetylase (LpxC), is crucial for the first hydrolysis step of the *N*-acetyl group in the biosynthesis of lipid A which anchors the lipopolysaccharide in the outer leaflet of Gram-negative bacteria. Diyne-based inhibitors of LpxC with a hydroxamic acid containing head group and various tail groups usually bearing primary hydroxyl group functionality can be potently bactericidal in vitro in Gram-negative pathogens as with wild type and clinical strains of *Pseudomonas aeruginosa*, although cardiovascular safety issues were evident in vivo (rat assay) (Cohen et al. 2019). From the same group, small molecules which have binding affinity for the lipid A biosynthetic enzymes LpxA and/or LpxD in *Pseudomonas aeruginoa* have also been identified (Kroeck et al. 2019) from physical studies on the enzymes and further structural modification work is suggested as a worthwhile future investigation path. Such dual binding could also reduce the rate of resistance development.

Other compounds have also been shown to be bactericidal in *Escherichia coli* and multidrug-resistant strains through inhibiting the lipopolysaccharide transporter MsbA, an inner membrane ATP binding cassette (ABC) transporter which mediates the critical initial stage in trafficking of LPS to the outer membrane (Alexander et al. 2018; Vetterli et al. 2018). Small molecules targeting MsbA include the potent and selective quinoline inhibitors G592 and G907 (Fig. 5.2a and b) (Alexander et al. 2018; Ho et al. 2018). Intriguingly, the bactericidal compound G907 was shown to have a dual-mode of inhibition of MsbA. One mode involved trapping the MsbA with LPS bound in an inner facing conformation by binding in a conserved transmembrane pocket. The second mode involved an allosteric binding modality which uncoupled the nucleotide-binding domains. This appears to be another promising approach to new antibacterials and it is also further confirmation that small molecules can modify bacterial transporter function.

Some elements of similarity in structure between a known LpxA and LpxD binder identified by Kroeck et al. (2019) (a reduced 2*H*-1,4-benzoxazine derivative with a terminal carboxylic acid group in one of the substituents) and G907 (Ho et al.

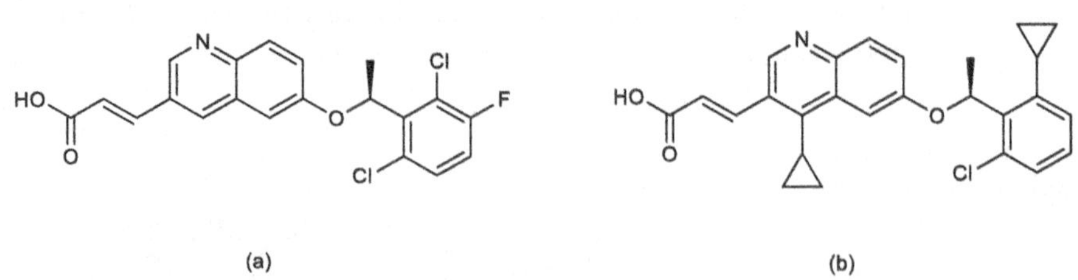

Fig. 5.2 Structures of the MsbA transporter inhibitors G592 (**a**) and G907 (**b**)

2018) also suggests it might be feasible to develop hybrids targeting both MsbA and LpxA and/or LpxD. Also exploring commonalities in design for lipid A biosynthesis enzyme inhibitors and then lipid A transporters would be a good approach in view of substrate similarities.

5.3 Search for Different Chemical Structure Types

Identifying and accessing unprecedented molecular scaffolds for incorporation in new hybrid and prodrug designs is still very relevant. Exploration at the same time of novel and unusual functional groups, such as the pentafluorosulfanyl group (referred to in Chap. 3, Sect. 3.3.4.1) can further extend the boundaries of new structural space. The new scaffolds could come from many sources and it will be increasingly important to be innovative in the search thinking well outside the norm and proposing unusual skeletal atoms and possible new functional groups incorporating elements from many parts of the Periodic Table. New directions in diversity-oriented synthesis are also likely to reveal a plethora of small molecules with increasingly complex molecular structural arrays and new bioactivities (Gerry and Schreiber 2020).

As a number of antibacterial natural products have shown at least dual modes of activity it would not be unreasonable to look further for new natural sources based on identifying silent operons in bacteria and other competitive microorganisms which if activated might express new multiply active compounds (Cully 2018; Ahmad et al. 2020). There is still a vast scope for research in this area and only some suggestions are touched on here. Exemplary strategies to establish viable platforms for antibacterial discovery, including previously non-accessible natural origins, have been thoroughly described (Lewis 2013; Brüssow 2017; Lok 2015). Again there is a need though to think well outside the square here and broaden the potential for serendipitous discoveries, including further 'mining the microbial dark matter' for new antibiotics (Lok 2015).

In the environment, many organisms, including microorganism, produce antibiotics as a self-protection mechanism to control or ward off attack by bacteria. Identifying such producers and the nature of the compounds and how they might act is an ongoing and exciting field of endeavour. Biorational approaches are particularly powerful for such discovery with one example being, among a number of others, the identification of the natural product tyriverdin, as a potent bacteriostatic agent isolated from the egg masses of the marine mollusc *Dicathais orbita* (Benkendorff 2013; Benkendorff et al. 2000).

It may also be beneficial to look again at seemingly disparate areas at the microbiological level in biorational searches. For example, looking for bacteria-protozoa defensive interactions. There are a number of reports of protozoa producing antibacterials including the perylenequinonoid (hypericin-related) derivative blepharismin from the ciliated protozoan *Blepharisma japonicum* (Pant et al. 1997). Blepharismin displayed good activity against MRSA but this activity was higher with photoactivation suggesting the possible involvement of singlet oxygen-derived reactive species in this case. The protozoal toxin, climacostol, an alkenyl resorcinol derivative (Petrelli et al. 2012), and the protozoal secondary metabolite, stentorin, also a hypericin-like pigment (Lobban et al. 2007), also have antibacterial activity. Blepharismins have been shown to be used for chemical defense in predator–prey interactions in other *Blepharisma* species (Buonanno et al. 2017).

With the biorational approach one might look more extensively for the production of antibacterials where any microbial competitors might be a severe threat because of the already tenuous viability under the conditions. Kown examples include the sideromycins but more are likely to be found. Analogues of the acyldepsipeptides (ADEPs) produced by bacteria have shown very promising antibacterial activity by dysregulating the highly conserved ClP serine protease enzyme involved in turnover of a range of bacterial proteins. Two of the natural ADEPs are the cyclic peptides enopeptin A and B (Carney et al. 2014).

Targeting non-multiplying bacteria is a new approach to the discovery of antibiotics as disclosed by Hu et al. (2010b) and the 1*H*-pyrrolo[3,2-*c*]quinoline derivative (HT61; Fig. 5.3) is a small molecule noted to have activity against biofilms of *Staphylococcus aureus* (Frapwell et al. 2020). It is also effective against other bacteria both alone or in combination.

Continuing to look for more antibacterials produced by bacteria in unusual places is likely to lead to new multi-targeting compounds. These may arise for example from extremophiles (Giddings and Newman 2015) as in the recent discovery of the salinipeptins from a salt-tolerant *Streptomyces* sp. from the Great Salt Lake in Utah, USA (Shang et al. 2019). These compounds contain D-amino acids, belong to the rare linearidin group, and have some antibacterial activity. Salinipeptin A was active against *Streptococcus pyogenes* in vitro, but not against some other Gram-positives or Gram-negatives.

Another unusual site for antibacterial discovery is the human biome as exemplified by the identification of lugdunin, a non-ribosomal cyclic peptide, derived from *Staphylococcus lugdunensis* in the nose. Lugdunin (Fig. 5.4) has an important thiazolidine ring embedded in the cyclic peptide, the nitrogen of which would be quite basic. Lugdunin shows potent activity against *Staphylococcus aureus* and other Gram-positive pathogens (Zipperer et al. 2016). Lugdunin shows potent activity against *Staphylococcus aureus* bacteria in this part of the biome. Development of resistance to lugdunin was not seen in *Staphylococcus aureus* over a period of 30 days

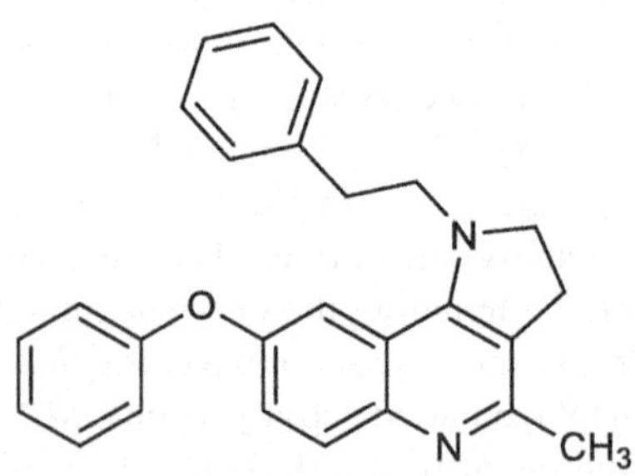

Fig. 5.3 Structure of the biofilm-targeting antibacterial HT61

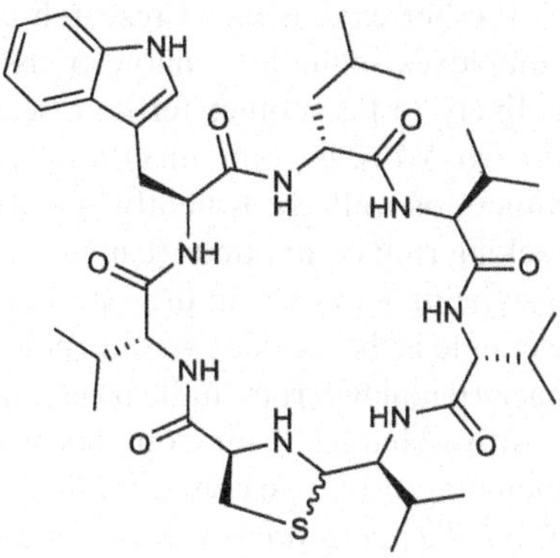

Fig. 5.4 Structure of the macrocyclic antibiotic lugdunin

from serial passaging of sub-inhibitory levels of the antibiotic. Intriguingly, when bacterial cells were exposed even to sub-MIC levels of lugdunin, incorporation of DNA, RNA, protein or cell-wall precursors ceased near concurrently implicating a fast deterioration in energy resources in the cell. Model and other studies suggest that the antibacterial activity of lugdunin in *Staphylococcus aureus* is associated with membrane potential nullifying effects through it's ability to translocate protons by acting as a carrier or possibly by channel formation (Schilling et al. 2019).

A very interesting paper by Chu et al. (2016) points the way to a new approach to antibiotic discovery based on a synthetic-bioinformatic approach, exemplified by the analysis of non-ribosomal peptide synthetase gene clusters from human-associated bacteria (human commensal and pathogenic bacteria). The antibacterial peptidic humimycins A and B (Fig. 5.5a and b) with a (3*S*)-3-hydroxymyristic acid acylated *N*-terminus were identified from this work. This hydroxy acid is a common fatty acid constituent of the lipid A part of bacterial lipopolysaccharides. The humimycins, which have been chemically synthesised, were found to inhibit lipid II flippase and potentiate β-lactam activity in MRSA in vivo.

Zhan et al. (2019) have also reported on peptides which are self-derived and which target methionine aminopeptidase in pathogenic bacteria and which have activity in a cell model against *Neisseria gonorrhoeae*.

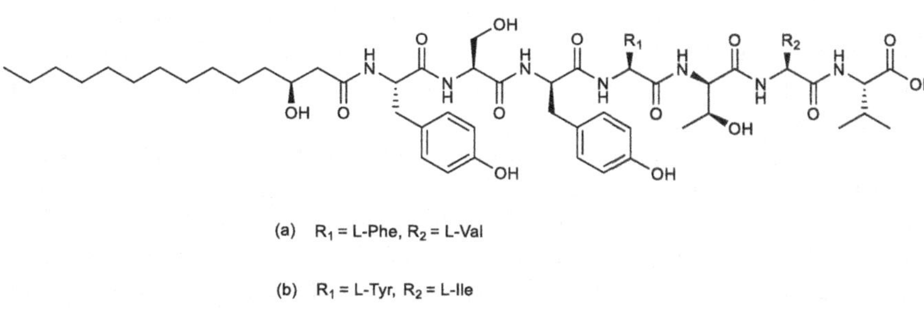

Fig. 5.5 Structure of humimycin A (**a**) and humimycin B (**b**)

Further expansion of research looking at the antibacterial actions of molecular complexes including charge transfer complexes with two (or more) components is likely to be another fertile research area, as long as such complexes stay intact in vivo. Work is going on currently in new drug discovery on non-covalently linked dimers, as well as covalently linked compounds, for protein target interactions. Interlocking ring compounds (catenanes) and rotaxanes and pseudo or quasi-systems are also being explored in drug design particularly antibacterial design. A good recent example is the work of Wu and colleagues who have described an intriguing strongly bactericidal heterometallic triangular necklace with Cu(I) and Pt(II) centres and nine positive charges. This complex was active against a range of bacterial pathogens including drug resistant strains of *Pseudomonas aeruginosa*, *Escherichia coli* and *Staphylococcus aureus*. Multi-targeting was evident with this complex resulting in DNA cleavage as well as disruption of the cell wall/plasma membrane (Wu et al. 2020).

5.3.1 In Silico *Advances*

The use of artificial intelligence (AI) technology by medicinal chemists to expedite the drug discovery process is advancing (Chen and Engkvist 2019) and further refinements are likely to lead to major developments in this area in the future including with respect to multiply active antibacterials. AI has the potential to come up with radically new designs for such agents.

Mapping biologically active chemical space is a continuing challenge but may be met in part by the use of AI as in the new platform called ASPIRE for pre-clinical drug discovery with the integration of automated synthetic chemistry with high throughput biology and artificial intelligence capabilities (Sittampalam et al. 2019). Planning syntheses will also be expedited using AI (Segler et al. 2018). The need to diversify by expanding the range of synthetic reactions used including developing new ones has been discussed by Brown and Boström (2016) and updated in 2018 by Boström et al. (2018). These broadened synthetic capabilities are likely to be required to implement novel multi-targeting antibacterial structural designs, as well as in the synthesis of more diverse chemical libraries for screening and then in post-screening optimization processes. As alluded to in different parts of this book the development of new reactions will also be assisted by thinking well outside the square in terms of what might be possible and involve advanced AI techniques in advancing the thoughts to specific reaction combinations.

The key to the future will revolve around synergistic interactions between human thinking and artificial intelligence across the full gamut of the antibacterial discovery and development paradigm including synthetic schemes, as well as toxicity and other ADME predictions (e.g., via the Centaur platform—W. van Hoorn, Exscientia and Centaur Chemist™; and comment by Mullard 2017; Luechtefeld et al. 2018). Awareness, though, of human biases in selecting appropriate algorithms in the AI process will be important and ways to reduce it might be drawn from outside the norm thinking (for such biases in chemical synthesis see Jia et al. 2019). AI also has the potential to go beyond molecular docking for drug mode of action insights when high quality data sets exist, which hopefully should be applicable in the targeted antibacterial area (Rodrigues et al. 2018).

5.4 New Modes of Action/New Targets

Together with the search for different molecular structures the search for new bacterial targets will continue to be of pressing importance. These two areas then interact in the ligand design process with deliberate multi-action design in mind. New target searching will necessarily involve cross disciplinary approaches perhaps in part via the computer-based platforms like the commercial Discuva platform developed by Summit Therapeutics Inc. This platform is aimed at the discovery of new bacterial targets and the development of new antibiotics to attack these targets. The company used this platform to identify a small molecule antibacterial SMT-571 with a new mechanism of action (associated with cell division) for drug resistant strains of the pathogenic bacterium *Neisseria gonorrhoeae* (Jacobsson et al. 2019).

The use of novel in-cell NMR techniques could herald new opportunities for drug design and discovery through a direct assessment of detailed protein-drug interactions in living bacterial cells. The NMR structure determination of a heavy metal binding protein overexpressed in *Escherichia coli* has been described (Sakakibara et al. 2009; Ikeya et al. 2016), as well as other work on the NMR-based determination of high resolution protein structures in living eukaryotic cells, including expressed proteins originating from bacterial sources such as the *Streptococcus* protein G B1 immunoglobulin-binding domain (a cell surface protein). The structure of this protein was the one determined most accurately under these physiological conditions (Tanaka et al. 2019). Further extensions of this work could give new insights for the design and assessment of selective antibacterials for example with three separate binding sites on the one target protein, if concentration and resolution problems can be overcome.

Recently multiple e-pharmacophore modelling (an e-pharmacophore is a structure-based energy optimised pharmacophore) has been used to identify novel inhibitors with good binding capability to two active sites on the bacterial 30S ribosomal subunit (Anju et al. 2019). Potentially this could be expanded to three sites or another target site affecting protein synthesis in addition to these two ribosomal subunit sites.

It could also be worth pursuing a wider assessment of the neat bioaffinity approach using native mass spectrometry (MS) methodology, in which the protein assembly is retained in their native states in the gas phase, as applied by Quinn and collaborators at Griffith University in Queensland, Australia (Vu et al. 2018) for potential protein targets from the malaria-causing protozoan, *Plasmodium falciparum*. This is a very interesting way to locate new hits for these targets. Some 96 natural products of low molecular weight were found as binding partners of 32 potential Plasmodial targets, and 79 of the natural products showed promising growth inhibitory activity in vitro against *Plasmodium falciparum.* Quinn and collaborators have also used native mass spectrometry in the bacterial domain applied to natural product fractions in which activity against *Mycobacterium tuberculosis* had been identified through high throughput screening. These fractions were then checked by native MS against a range of purified possible possible targets from *Mycobacterium tuberculosis.* Further analysis then uncovered a promising new natural product lead, altholactone, which binds, probably via irreversible covalent binding, to the mycobacterial protein Rv1466. Altholactone had moderate activity against *Mycobacterium tuberculosis* in vitro (Elnaas et al. 2020). Extensions of the methodology might embody looking for triple or higher order binding by ligands to one or more bacterial protein targets. Also three possible binding sites could be reasonably proposed through computer-based modelling and docking and then checked through bioaffinity mass spectrometry with the appropriate proteins. Alternatively one could increase the number of proteins and look for any binding ligand. Some bacterial targets are not proteins though and they would also need to be accessible in sufficient amount and purity for the mass spectrometric studies.

It has also been shown recently that the strategy named PROSPECT (primary screening of strains to prioritize expanded chemistry and targets) can be used to identify new hit compounds which in *Mycobacterium tuberculosis* (Mtb) inhibited a new target found to be the essential efflux pump EfpA. These hit compounds can then be developed or optimised into good leads. With PROSPECT, library compounds were screened against pools of engineered Mtb hypomorph strains, in which vital bacterial targets were greatly diminished. Specific activity-enriched and unbiased compound libraries were used to identify new inhibitors targeting DNA gyrase, mycolic acid biosynthesis, folate and tryptophan biosynthesis, RNA polymerase, as well as the EfpA efflux pump. Significantly, this target depletion-based strategy identified new inhibitors against targets that would not have been found using screening based on wild-type Mtb alone. Also, this approach should be applicable to other pathogenic bacterial strains, and potentially provide information on compounds with more than one mode of action which possibly would not be revealed by conventional screening methodology (Johnson et al. 2019).

Toxins and toxin inhibitors

Nullification by small molecules of bacterial endo- or exotoxins and their actions is a potentially rich area for further development of multi-active agents. Various steps could be targeted including their synthesis and release from bacteria, binding of ligands to key sites on the toxins required for activity, and antagonising the binding of toxins to host cell receptors. Bacterial toxin inhibitors have been developed based on multivalent scaffolds with multiple sites for interaction. These inhibitors are based on glycopolymers, glycodendrimers, or glycoclusters and are quite large molecules (Branson and Turnbull 2013). Monovalent and multivalent carbohydrate receptor-binding inhibitors of cholera toxin, the entero toxin secreted by *Vibrio cholera*, a pathogenic Gram-negative facultative anaerobe and cause of cholera, have also been described by Kumar and Turnbull (2018), although again these mostly tend to have quite high molecular weights. These compounds are capable of binding to a number of sites on the cholera toxin receptor.

Bivalent inhibitors of cholera toxin linked via a 1,2,3-triazole unit are discussed by Leaver et al. (2011), although from this work the strongest inhibitors of cholera toxin binding to its cell surface receptor were seen with monovalent analogues of lower molecular weight. These observations could help with trying to design more potent smaller molecule equivalents and with other targeting substituents.

Instead of antagonists binding to the toxin cell surface receptor, one might consider compounds which bind directly to the toxin and thus inhibit effective subsequent binding to the receptor. Clues to inform the design of such decoy toxin traps might be expected to come from a detailed study of decoy exosomes released by the host cell to protect against bacterial toxins as described by Keller et al. (2020). These decoy exosomes, which bind to the toxins, were shown to be active in vitro and in vivo. In the latter case they were shown to improve survival in MRSA-infected mice.

Virulence

Considerable scope for multi-targeting is predicted for antivirulence approaches to attack pathogenic bacteria and this active area has been covered in a recent review (Calvert et al. 2018). One example of this approach involves the inhibition of macrophage infectivity potentiators (Mips), which are proteins belonging to the FK506-binding family. They can be important for bacterial virulence as with the Mip from *Burkholderia pseudomallei,* the intracellular bacterial pathogen which causes meliodiosis (Begley et al. 2014). Mips play a role in the invasion of human macrophages, the white cell type of the immune system responsible, among other activities, for detecting, engulfing and digesting pathogens. Mips attenuate the effectiveness of the macrophage in its host protection role. The antibiotic Rapamycin can inhibit this Mip as can some pipecolic acid derivatives and a future endeavour may wish to look at ways in which this inhibitory activity could synergise with other activities.

A further example of antivirulence compounds are those developed to negatively impact on DsbA, the Gram-negative, periplasmic thiol-disulfide oxido/reduction system important for bacterial virulence (Heras et al. 2015). In view of this, inhibitors of this system are being investigated using a fragment based approach to identify a new structural class of benzofuran inhibitors of DsbA from *Escherichia coli* (Duncan et al. 2019). The strongest binders to DsbA were found to be 2-(6-phenoxy or 6-benzylbenzofuran-3-yl)acetic acid derivatives which bind in the hydrophobic groove of the enzyme close to the catalytic disulfide bond. Further optimization of these structures is indicated. The strongly oxidising DsbA is a member of the Dsb (disulfide bond) family of enzymes which catalyse critical intramolecular disulfide bond formation in peptides as they enter the periplasm of the cell. The disulfide bond formation occurs in a range of virulence factors and, significantly, without functional DsbA such bacteria have reduced virulence, heightened sensitivity to antibiotics and a reduced capacity to initiate infection (Heras et al. 2015).

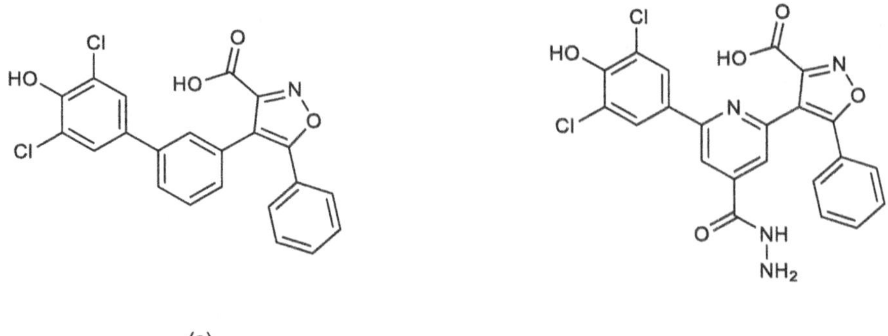

Fig. 5.6 Structure of an isoxazole inhibitor (**a**) of MptpB phosphatase enzyme and a suggested structure for an isoniazid-isoxazole hybrid (**b**)

Another very promising antivirulence approach is that focussed on inhibitors of the phosphatase enzyme MptpB, a secreted virulence factor that disrupts host-based antimicrobial activity for *Mycobacterium tuberculosis* in vivo. Small molecule inhibitors of this enzyme have been uncovered (Vickers et al. 2018). A structure-based design approach was used in this work and one of the potent isoxazole compounds developed is shown in the Fig. 5.6a. This orally bioavailable inhibitor ameliorated the infection burden in vivo (chronic and acute models in the guinea pig). This and related compounds also enhanced bactericidal efficacy of other key clinical antibiotics. A suggested chimeric hybrid structure (Fig. 5.6b) of the prodrug isoniazid and an isoxazole-type MptpB inhibitor providing multi-targeting capabilities might be worth exploring here. Isoniazid is activated by the bifunctional catalase-peroxidase enzyme KatG to give the active isonicotinoyl radical and this should still be possible to give a disubstituted analogous radical with the hybrid (Fig. 5.6b). From a visual assessment of the model of componds docked in the active site of the bacterial enzyme in the Vickers paper (2018), replacement of the central phenyl ring by a pyridyl ring would not seem to be a problematic change although the positioning of the hydrazide may be in terms of accommodating binding in the key primary phosphate binding pocket (P1). Computer-based docking would need to be undertaken to assess this.

A different way to counter virulence is through the development of pilicides, for example substituted thiazolo[3,2-*a*]pyridones (bicyclic 2-pyridones), that interfere with pili formation, Åberg and Almqvist (2007). Pili are virulence-associated organelles which are important for the attachment of bacteria to host cells and thus if their growth is compromised it affords a way to reduce virulence. These pilicides attenuate the conserved chaperone-usher pathway which is crucial for the complex multi-protein assembly process to construct pili. For a later review on chaperone-usher (CU) function and assembly and including further information on inhibitors of pilus-mediated adhesion and small molecules which can interfere with the biogenesis of pili see Psonis and Thanassi (2019).

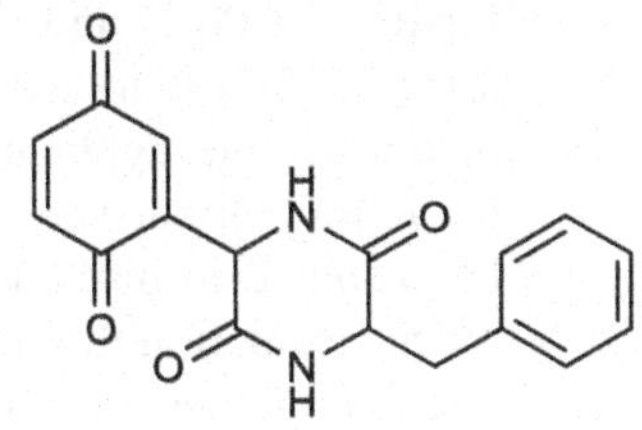

Fig. 5.7 A diketopiperazine inhibitor of quorum sensor synthase CepI

Quorum sensing

Molecular targets external to those within bacteria, such as those involved in the quorum sensing sphere, are a general area of increasing interest as highlighted in the review by Monserrat-Martinez et al. (2019). Quorum sensing is a complex but vital communication paradigm for bacteria and disruption of the communication is an ongoing area of research (Haque et al. 2018). A wide range of approaches for disrupting quorum sensing in bacteria can be considered for future applicability including (i–iii):

(i) inhibition of enzymes which synthesise the signalling agents looking at both substrate-based and transition state-based design, although inherent resistance processes to overcome enzyme inhibitors may weaken this approach.

Diketopiperazine-based inhibition of quorum sensor synthase CepI in the Gram-negative human pathogen *Burkholderia cenocepacia* in vitro has been described (Buroni et al. 2018). From computer-assisted studies and a homology model, together with site-directed mutagenesis work, it was suggested that the diketopiperazine (Fig. 5.7) exerted its inhibitory effects through interaction with a flexible loop implicated in the recognition aand stabilization of the enzyme substrate *S*-adenosylmethionine. This compound did not have any direct antibacterial activity but it was shown that survival of the *Burkholderia cenocepacia*-infected nematode *Caenorhabditis elegans* was increased indicative of reduced virulence of this strain of the bacterium in vivo. *Caenorhabditis elegans* is a useful living intermediary for the discovery of anti-infectives in general (Arvanitis et al. 2013).

A good recent review with regard to interfering with the *Pseudomonas* quinolone signal quorum sensing system with small molecule interventions is presented by Schütz and Empting (2018). In this review small molecules which can interfere with the biosynthesis of the sensors as well as modulating receptor interactions, amongst others, are discussed and includes useful detail on computer-modelled binding modes. Within the quinolone system the signal molecules involved, 2-heptyl-3-hydroxy-quinolin-4(1*H*)-one (PQS) and the immediate biosynthetic precursor for it without the 3-hydroxy group (HHQ), induce the transcription of a number of genes including their own biosynthesis enzymes (PqsABCDE). These enzymes are involved in a cascade sequence to synthesise PQS from anthranilic acid. Various small molecule inhibitors of these enzymes, including promising dual targeting inhibitors of PqsBC and

PqsR or of PqsD and PqsR, are discussed. Further studies on possibly triple targeting inhibitors would be of great interest.

(ii) increasing the application of mobile 'mini' receptors to strongly complex the extra-cellular sensor. Such 'Pied piper' or 'sponge' therapeutics could be designed to lure and strongly bind quorum sensing molecules including by encapsulation through multiple non-bonded interactions with one or more ligand molecules resulting in a Russian-doll type complex.

(iii) inhibition of sensor-receptor binding through a competitive or non-competitive antagonist. Small molecule autoinducer analogues capable of modulating the SdiA quorum sensing receptor in *Salmonella enterica* serovar Typhimurium via agonism or antagonism have been reported by Styles and Blackwell (2018). Nizalapur et al. (2016) have also reported on a number of structurally novel inhibitors of bacterial quorum sensing based on *N*-arylisatin-based-glyoxamide derivatives which are accessible synthetically via the reaction of *N*-aryl isatins with cyclic and acyclic amines resulting in isatin ring opening from attack at the C2 lactam carbonyl group. One of these derivatives, an ethyl glycinate-substituted glyoxamide, exhibited an interesting activity profile with moderate quorum sensing inhibitory activity against a LuxR-expressing *Escherichia coli* strain and also against a strain of *Pseudomonas aeruginosa* expressing LasR receptor. Computer-based docking studies with LasR protein indicated the important role of hydrogen bonding interactions in the docking of this compound with the highest quorum sensing inhibitory (QSI) activity.

Biswas et al. (2018) also studied some fimbrolide analogues as quorum sensing inhibitors in a *Pseudomonas aeruginosa*-based assay, and interestingly one of the compounds exhibited quorum sensing inhibitory activity as well as bactericidal activity suggesting at least dual activity. This offers further potential for extension to triply active hybrids.

It should be noted here, though, that although the approach of targeting the QS system is thought to be one which will not lead to a bacterial resistance response, this is not necessarily the case as demonstrated by Maeda et al. (2012). This work is referenced in the useful review by Arvanitis et al. (2013) which includes coverage on using *Pseudomonas aeruginosa* in *Caenorhabditis elegans* as the in vivo model for assessing the effects of exposure to an agent that targets quorum sensing.

Selective modulation of bacterial kinases

Bacterial kinases are another bacterial signalling system which can be targeted. Some inhibitors of penicillin-binding-protein and serine/threonine kinase-associated (PASTA) kinases are relevant as these kinases have a role in mediating resistance to β–lactam antibiotics. These dual enzyme inhibitors, for example GSK690693 (Fig. 5.8), are imidazopyridine aminofurazans (Schaenzer et al. 2017). GSK690693 has been shown to sensitize the bacterial pathogen *Listeria monocytogenes* to the β-lactam antibiotic ceftriaxone. Extensions here to involve multi-targeting suggest a

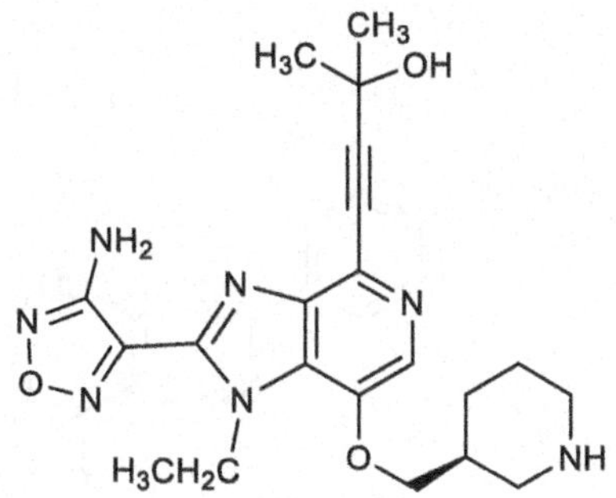

Fig. 5.8 Structure of the dual enzyme inhibitor GSK690693

combination approach with inclusion of a β-lactamase inhibitor or a monobactam in the mix might be beneficial.

Bacterial QseC is a histidine sensor kinase and inhibitors of this kinase have also attracted attention—for example the mixed thiourea-sulfonamide LED209 (Fig. 5.9a). This compound inhibits the binding of signals to the membrane histidine sensor kinase, Qsec, which responds to adrenergic signalling molecules in the host and to bacterial signalling molecules to then stimulate virulence factor expression. While the compound does not inhibit growth of bacteria it has been shown to inhibit the virulence of a number pathogens in vitro and in vivo in animal models. These kinases are promising targets for selective actions as they are not present in mammals (Rasko et al. 2008). A rotationally more restricted analogue could also be of interest and one way to achieve this might be to incorporate part of the thiourea moiety in a ring as in a benzothiazole derivative or a benzimidazole derivative (Fig. 5.9b). The benzothiazole group has also been used for example as a useful bioisostere for a urea moiety in the anthelmintic benzoylphenylureas (Baell 2021).

Other enzyme targets

Synthesised derivatives of the arylomycin natural products have revealed a new antibacterial, G0775 (Fig. 5.10), active against Gram-negative pathogens, including drug resistant ones, which inhibits the new antibacterial target LepB, a bacterial

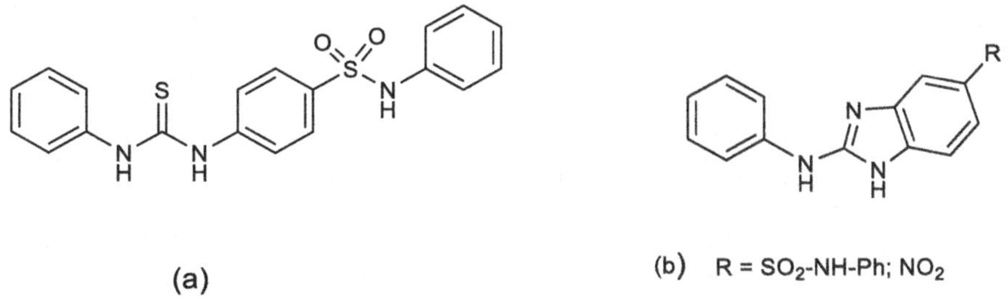

Fig. 5.9 Structure of the bacterial histidine kinase inhibitor LED209 (**a**) and suggested benzimidazole analogues (**b**)

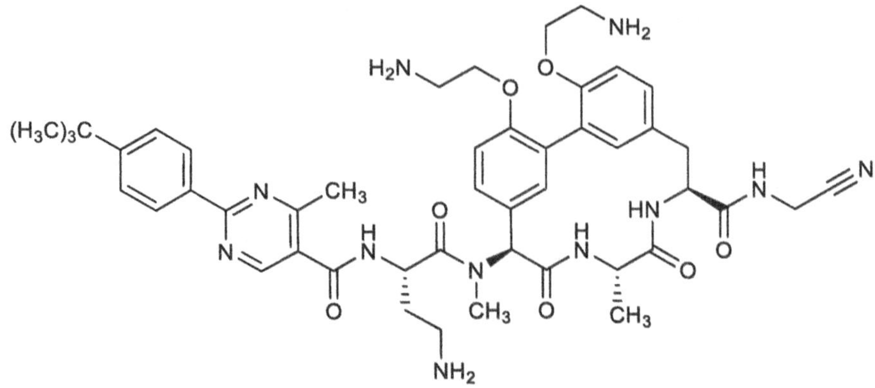

Fig. 5.10 Structure of the LepB inhibitor G0775

type I signal peptidase. LepB is an essential inner membrane enzyme intimately involved with the protein export machinery (Smith et al. 2018). This compound resulted from efforts to optimize the antibacterial activity of the naturally occurring weak and restricted spectrum antibacterials, the arylomycins for which the structure of arylomycin A-C_{16} is shown (Fig. 5.11). G0775 retains the macrocylic ring of arylomycin A-C_{16} but has a number of significant changes in the attached substituents including exposed *N*-cyanomethyl functionality which mediates covalent binding to a lysine residue through formation of an amidine unit in the substrate binding site. However it is not clear how this might be developed into synergistic activity with the involvement of other binding sites in view of the irreversible nature of the covalent binding. Still it is an exciting new antibacterial.

Inhibitors of another bacterial membrane protein MraY (translocase I), an important pan-bacterial enzyme concerned with the peptidoglycan aspects of cell wall biosynthesis, obviate the formation of lipid I a vital intermediate in the construction of the bacterial cell wall. This target for new antibacterials has been reviewed by Fer et al. (2018) while Koppermann et al. (2018) discuss target interactions of the

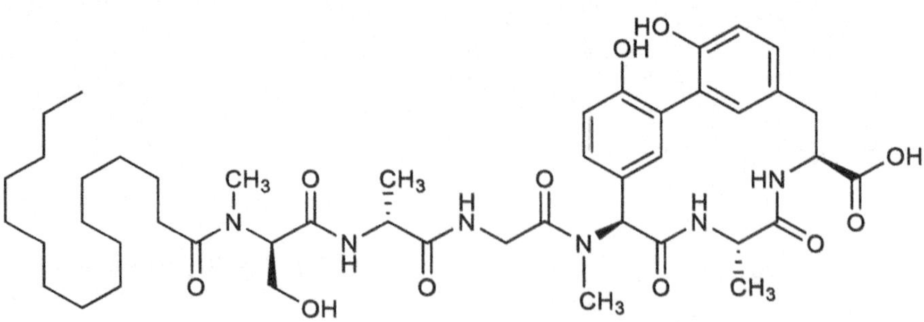

Fig. 5.11 Structure of arylomycin A-C_{16}

muramycin nucleoside-based antibiotics as a foundation for new optimized analogue development. There would seem to considerable potential here for hybrid or prodrug design to include cephalosporin or similar units. The development of such hybrids or prodrugs could usefully be based on the sites identified for MraY inhibition by Mashalidis et al. (2019) as 'druggable hotspots' on the enzyme. Some six such hotspots have been identified as a result of extensive co-crystallization studies with known types of inhibitors and subsequent X-ray crystallographic analysis, apart from the essential common uridine binding site. Perhaps a fluoroquinolone could also be modified to bind to one such hotspot while still retaining the gyrase and topoisomerase IV inhibitory activity.

As introduced in Chap. 1 (Sect. 1.3.4), the bacterial survival strategy of programmed cell death in bacterial communities (Peeters and de Jonge 2018) may also be a potential bacterial vulnerability in which unprogrammed, or premature, cell death might be initiated via small molecule interventions and enzyme interactions, which could be detrimental to the bacterial community public good (Tanouchi et al. 2013). Careful structural design work would be required for such molecules with appropriate enzyme multi-targeting. A potential issue may be that perhaps there are endogenous molecules in bacteria which sense premature bacterial cell death and which then initiate other actions to locate and nullify the problem. If such compounds were identified in the future then interference with the actions of any such molecules could be a fruitful new avenue for antibacterial research.

Other protein targets

Bacterial chaperone system

Bacterial chaperones afford other opportunities for antibacterial interventions and the use of triply active compounds to effectively target these chaperones such as HSP70, known as DnaK, and it's two partner proteins (Alix 2013). Targeting bacterial chaperones in the bacterial proteostasis system for the treatment of tuberculosis has been reported by Lupoli et al. (2018) and the prospects for more such selective targeting developments are likely to be good.

The chaperonins GroEL and BroES are critical for bacterial growth and small molecule inhibitors are also being actively pursued. These inhibitors have been identified from large library screening (Abdeen et al. 2016, 2018; Stevens et al. 2019).

An interesting development related to the bacterial chaperone context is the discovery, from a screen of *Photorhabdus* symbionts of nematode microbiomes (see also Sect. 5.3 for other approaches), of the antibiotic darobactin, which is selectively toxic to Gram-negative pathogens in vitro and in mouse models of infection (Imai et al. 2019; and useful explanatory comment on the work by York 2020). Darobactin (Fig. 5.12) is a modified hexapeptide with two unusual embedded macrocyclic components and it acts through interference with BamA a chaperone and translocator which folds outer membrane proteins. While darobactin-resistant mutant strains of *Escherichia coli* could be induced in laboratory culture, this result reinforces the

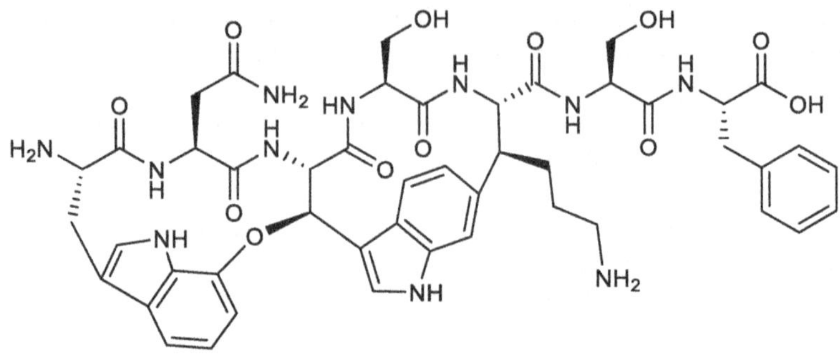

Fig. 5.12 Structure of the new antibiotic darobactin

validity of the chaperone inhibitor approach for further development and incorporation into multiactive antibacterial designs. It is also of interest to note that two other glycopeptide antibiotics found from an evolution-guided discovery approach, namely the known compound complestatin and the previously unknown relative named corbomycin, both contain some similar structural elements to darobactin including the unusual 3,5-disubstituted indolic linking unit; the former two antibiotics affect remodelling of the cell wall during growth through binding to peptidoglycan and inhibiting peptidoglycan hydrolases (autolysins) crucial for the remodelling process (Culp et al. 2020).

In other important studies it has been shown that some synthetic chimeric peptidomimetics have potent broad spectrum bactericidal activity against Gram-negative bacteria. These chimeric molecules contain two linked macrocyclic rings one being a β-hairpin peptide unit related to murepavadin and the other related to the macrocycles found in the polymyxin natural product family. Two modes of action seem to be involved one being binding to lipopolysaccharide and the other binding to BamA in the β–barrel folding complex (BAM) (Luther et al. 2019). These compounds are thus dual action hybrids and one optimized analogue is being taken on to clinical trials. Similarly the compound MRL-494 (Fig. 5.13) seems to inhibit BamA directly or possibly indirectly through targeting the outer membrane in Gram-negative bacteria (Hart et al. 2019) and summarised neatly by Sousa (2019).

Bacterial protein transport systems.

Targeting specific transporters in the outer membrane of Gram-negative bacteria is a growing area and future possibilities includ inhibition of these transporters as one arm of multi-action designs. Murepavadin is a *Pseudomonas*-specific macrocycylic peptide containing fourteen amino acids making up the ring structure, which targets transport of lipopolysaccharide (LPS) in the Gram-negative bacterium *Pseudomonas aeruginosa* (Sader et al. 2018). Murepavadin binds to LPS transport protein D (LptD) involved in the assembly of LPS in the outer leaflet of the outer membrane. This peptide was the first example of a compound class which targets an outer membrane protein. It is highly active in vitro and also in a mouse model of *Pseudomonas*

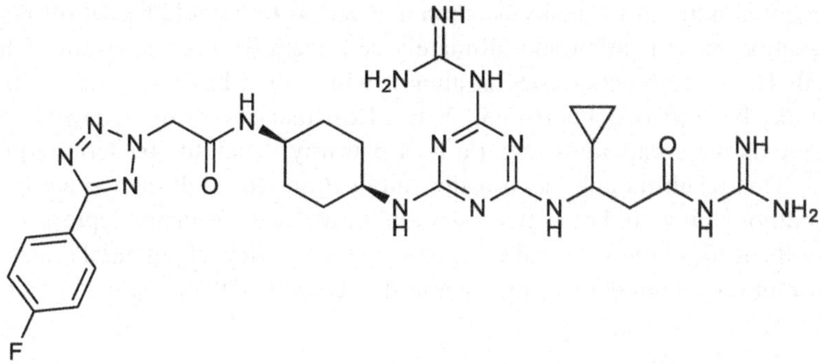

Fig. 5.13 Structural representation of MRL-494, a Gram-negative antibacterial

aeruginosa-induced septicaemia (Srinivas et al. 2010). Murepavadin should provide a good template for the development of other perhaps smaller molecule structures but ones which incorporate the key elements for transporter inhibition as well as pharmacophores for other target interactions.

Inhibition of protein–protein interaction in Gram–negative bacteria with thanatin, a 21 residue insect defense peptide, is a novel mechanism of action. In the recent review by Robinson (2019) folded synthetic peptides and other molecules targeting outer membrane complexes in Gram-negative bacteria, and the mode of action of thanatin interacting with the Lpt complex in *Escherichia coli*, is discussed.

Another related bacterial transporter, LolCDE, which is essential for the transpositioning of lipoprotein in Gram-negative bacteria, is inhibited by a pyrrolopyrimidinedione derivative (GO507, Fig. 5.14). Mutations were shown to confer significant resistance to this derivative suggesting that resistance will be a problem if antibacterial use is contemplated in the future. It could be a useful chemical probe, however, in view of its selective binding properties in the lipoprotein (Nickerson et al. 2018).

The modification of cholesterol transport within the cell is a further transport avenue worth pursuing. Alteration of this transport by small molecule functional inhibitors of the acid sphingomyelinase (ASMA)-ceramide system, such as the antidepressant desipramine, has been reported by Cockburn et al. (2019). Cholesterol

Fig. 5.14 Structure of the pyrrolopyrimidinedione derivative GO507

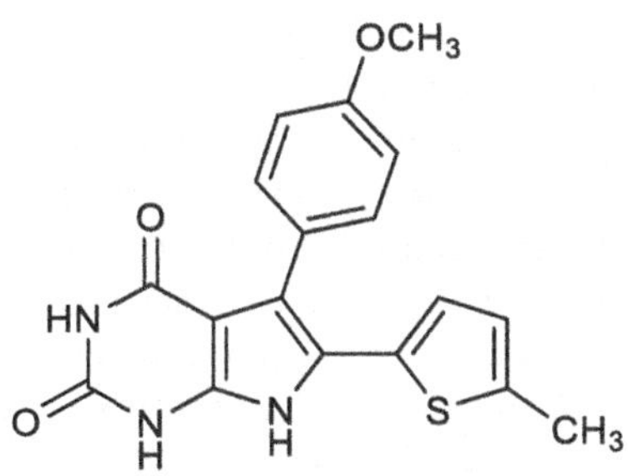

is an essential nutrient for many bacteria and access to it could be cut off resulting in a cessation of replication and ultimately cell death for the bacterium within the host cell. Host cell biochemistry is altered which then has a negative impact on intracellular bacteria (Cockburn et al. 2019). Resistance is considered to be unlikely to develop as the compounds target a host pathway which the bacteria require for survival. This result though has broader implications for multi-action antibacterial design, although it would be better to try and move away from antidepressant structural motifs in any composite multi-targeting design in view of potential issues which may arise in giving antidepressants to non-depressed healthy people.

5.5 DNA and RNA Level Modulation

Modulation of the DNA-RNA sphere in bacteria continues to be an important area from the perspective of new antibacterials including multi-targeting approaches. This momentum is likely to continue in the future and only some examples are indicated in this section.

One key area involves inhibition of DNA replication. Inhibition or suppression of DNA replication via interference with the sliding clamp (β-clamp) as the target for the antibacterial activity of some non-steroidal anti-inflammatory drugs as well as other compound types has great promise for the development of new antibacterials (Yin et al. 2014; Altieri and Kelman 2018). Additionally, detailed molecular understanding of the intricate bacterial DNA replication processes from cutting edge observational techniques, particularly single molecule techniques (Robinson and van Oijen 2013), will continue to aid the design process for specific inhibitors.

A further target for antibacterials involves counteracting bacterial DNA repair through small molecule interference with DNA repair proteins that act on Holliday junctions which are vital intermediates in many repair pathways. From a compound library screen, Rideout et al. (2011) discovered a number of interesting disubstituted pyrrolidine derivatives incorporating 5-membered ring cyclic guanidine units in each of the substituents. These compounds were shown to interfere with the resolution of Holliday junctions in vitro and were also potent growth inhibitors of both Gram-negative and Gram-positive bacteria, particularly the latter. Earlier work by Gunderson and Segall had identified Holliday junction-trapping hexapeptides (Gunderson and Segall 2006) but these are quite large molecules.

Transcriptional regulators have also been identified as important targets for different antibacterial approaches (González et al. 2018). Another potential avenue of attack is to consider epigenetic drug targets in bacteria arguing by inference from the insightful work of Baell and colleagues (Baell et al. 2018) who described small molecule inhibitors of particular histone acetyltransferases that induce senescence and arrest the growth of tumours; in effect the cells go to 'sleep'. Their most active compound was the aroylsulfonohydrazide WM-8014 (Fig. 5.15). Different targets with a similar function seem possible in bacteria (Hu et al. 2010a).

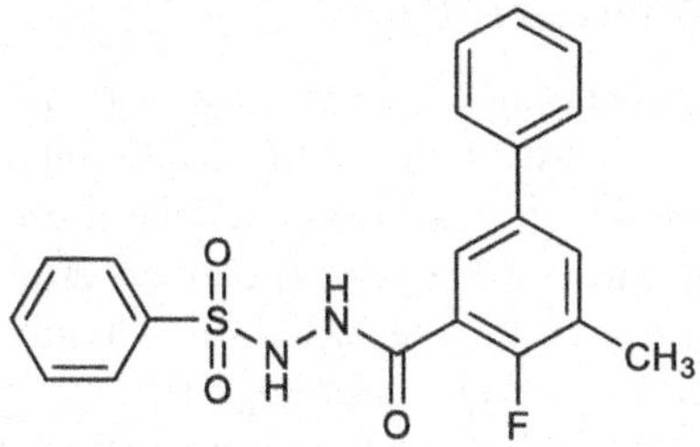

Fig. 5.15 Histone acetyltransferase inhibitor WM-8014

In the RNA sphere, there has been some work on RNA guided nucleases (Silver 2014) and riboswitches, the mRNA structures which regulate gene expression in bacteria. Riboswitches have been assessed as antibacterial drug targets (Blount and Breaker 2006) and dual-targeting of two riboswitches at the same time via the synthetic derivative ribocil-C and the natural product roseoflavin has been reported (Krajewski et al. 2017). Inhibitor design protocols for RNA polymerase and tRNAs are also likely to continue to be fruitful areas for targeted antibacterial research as noted in Sects. 5.2.1 and 5.2.2 in this chapter.

5.6 Proteins and Antibacterials

Protein degradation

There is much interest in drug-mediated protein degradation for potential therapeutic purposes as an alternative to inhibiting protein targets. Degradation of proteins selectively is being studied in human cells by proteolysis targeting chimeras (PROTACS). These chimeras can be formed in cells by 'click' reactions and they then trigger steps to induce selected protein degradation (Gu et al. 2018; Lebraud et al. 2016). Other studies indicate that the techniques may be applied to develop such chimeras for selective destruction of bacterial proteins. Targeted protein degradation has been shown to be possible with the antitubercular drug pyrazinamide, which can act as a promoter to trigger destruction of its target (Gopal and Dick 2020). Whether this approach can be extended to a range of other bacterial targets awaits further work but it has very significant potential.

Disarming of host defence proteins by bacterial proteins

Another interesting new target possibility is presented by bacterial protein-based effectors which interfere with host ubiquitin and ubiquitin-like proteins. Perhaps one or more actions of any new antibacterials could involve hindering these effectors thus leaving the bacteria more open to the host defence mechanisms (Ribet and Cossart 2018). Pathogenic bacteria lack such protein systems, but interestingly, *Bacteriodes fragilis* in the gut microbiome does seem to have a ubiquitin homologue and secretes it to implement intraspecies antagonism (Ribet and Cossart 2018).

Bacterial antigens

Circulating bacterial antigens (Nuti et al. 2011) also afford possible targets for therapies involving minimalised antibodies as pharmaceuticals (Rader 2015). Further studies in this area cover the use of synthetic immunotherapeutics against Gram-negative pathogens (Feigman et al. 2018). The design idea in this work was to use a conjugate based on polymixin B as the bacterium surface targeting agent and an antigenic epitope that can recruit antibodies found in human serum, with ultimate bacterial cell destruction. With an appropriate lipid side chain the conjugate itself can also be antibacterial against Gram-negative bacterial pathogens. So basically a two-pronged activity approach is used but extension to a three-pronged attack might increase potency even further while selectively targeting the problematic Gram-negatives.

5.7 Concluding Remarks

While various non-small molecule approaches to control bacterial pathogens are on-going including lysins, probiotics, phages, immune stimulation and vaccines, the emphasis in this book has been on the small molecule multi-targeting/activity design side. However, although current approaches to multi-activity have been productive, there is an urgent need to broaden the scope by thinking outside the square, including 'outside the bug' (Monserrat-Martinez et al. 2019), and considering the forest as well as the trees.

This raises a more fundamental question in just how does one think outside the square? How do you know you are outside? Or how does one 'imagine the unimaginable' as noted by Professor Ben Feringa at the end of his Nobel Prize Lecture (Feringa 2017). Is it all just serendipitous or can the likelihood of serendipitous discoveries be assisted by the nature of the journey, the paths one is on, or is something totally different required? Perhaps some pointers to answers can be taken from neural network representations and the formalisms or algorithms involved in neural network based artificial intelligence.

Also it is suggested that other answers are likely to come from looking carefully to nature again with new eyes and questions. Perhaps questions along the lines of: How is antibiotic resistance controlled in nature of which we are an integral part? Why is resistance not induced by antibiotics in their natural settings or is it? How do soil bacteria or microbiome bacteria (or other bacteria in competitive environments) overcome resistance to their secreted antibiotics to control other competing bacteria? Possibly there is a hint here based on recent work on a novel alkaliphilic *Streptomyces* species (sp. *myrophorea*, isolate McG1) from a Northern Ireland soil sample with activity against ESKAPE pathogens. The *Streptomyces* strain was resistant to 28 out of 36 clinical antibiotics and from *in silico* based gene analysis had many secondary metabolite and toxicity resistance gene clusters as well as a number of antibiotic resistance genes possibly related to antibiotic production (Terra et al. 2018). Interestingly, the alkaline/radon soil was investigated because of its ethnopharmacological use as a cure for infections. Other hints are very likely to come from, amongst other areas, extensions to work on how the gut microbiome is modulated (Elgart and Soen 2018), and the identification of new antibacterial compounds from nature using screening methods which employ antibiotic resisistance as a possible filtering criterion. This is seemingly contradictory but it can yield higher hit rates (Thaker et al. 2013).

It will be increasingly important to try and be fundamentally different in future antibacterial research, and not so much more of the same. This will not be easy to do. Is it possible to make a quantum leap rather than progressive small jumps as largely suggested in this book, although some attempts have been made to try and include some more radical suggestions for research. Quantum leaps, though, will hinge on new chemistry and new biology using compounds not considered today and which might engender new interactions with biological targets and be multiply and selectively active. One will still be limited in a sense by nature since the targets available will be those in the natural world like proteins, nucleic acids, fatty acids, or polysaccharides. But perhaps in future therapies these can first be made 'unnatural' to allow for radically new interactions with other small molecules? Also one wonders if new ways of destroying pathogenic bacterial cells from within can be devised possibly using light-mediated in-cell free radical polymerisation to produce toxic polymeric assemblages. Recent work has established that light-mediated free radical polymerisation was possible in HeLa cells using a biocompatible initiator and various monomers while maintaining cell viability (Geng et al. 2019). With pathogenic bacteria though the goal would be ultimate non-viability.

In the realisation and translation of any new or radical proposals multidisciplinary participation and collaboration will continue to be vital including collaborations between academia, research institutions and pharma (large, medium and small). It is unfortunate, however, that a number of big pharma have withdrawn from this antibacterial field like Novartis in 2018 (Megget 2018) or are showing declining interest. Another company Allergan is also divesting antibiotic assets and biotechs are not doing too well in this space (Shlaes 2018). Clear monetary reward incentives for antibiotic development are needed and continuing support through WHO, and international Trusts, Funds, Alliances, and Foundations will be increasingly important in the future (Wanted: a reward for antibiotic development 2018).

Much will always remain to be done and new avenues followed. The challenges of drug-resistant bacterial infections are great but the likely health benefits for the global community will hopefully provide the stimulus to never give up in the search for counter measures.

References

Abdeen S, Salim N, Mammadova N et al. (2016) GroEL/ES inhibitors as potential antibiotics. Bioorg Med Chem Lett 26(13):3127–3134

Abdeen S, Kunkle T, Salim N et al. (2018) Sulfonamido-2-arylbenzoxazole GroEL/ES inhibitors are potent antibacterials against methicillin-resistant *Staphylococcus aureus* (MRSA). J Med Chem 61(16):7345–7357

Åberg V, Almqvist F (2007) Pilicides-small molecules targeting bacterial virulence. Org Biomol Chem 5 (12):1827–1834

Ahmad I, Ahmad S, Rumbaugh KP (eds) (2020) Antibacterial drug discovery to combat MDR: Natural compounds, nanotechnology and novel synthetic sources. Springer Verlag, Singapore

Alexander MK, Miu A, Oh A et al. (2018) Disrupting Gram-negative bacterial outer membrane biosynthesis through inhibition of the lipopolysaccharide transporter MsbA. Antimicrob Agents Chemother 62 (11):e01142–01118

Alix J-H (2013) Targeting HSP70 to fight cancer and bad bugs: One and the same battle? In: Gualerzi CO, Brandi L, Fabbretti A et al. (eds) Antibiotics: Targets, mechanisms and resistance. Wiley-VCH Verlag GmbH, Weinheim, pp 525–538

Altieri AS, Kelman Z (2018) DNA sliding clamps as therapeutic targets. Front Microbiol 5: Article 87

Anju CP, Subhramanian S, Sizochenko N et al. (2019) Multiple e-Pharmacophore modeling to identify a single molecule that could target both streptomycin and paromomycin binding sites for 30S ribosomal subunit inhibition. J Biomol Struct Dyn 37 (6):1582–1596

Arvanitis M, Glavis-Bloom J, Mylonakis E (2013) *C. elegans* for anti-infective discovery. Curr Opin Pharmacol 13 (5):769–774

Baell JB, Leaver DJ, Hermans SJ et al. (2018) Inhibitors of histone acetyltransferases KAT6A/B induce senescence and arrest tumour growth. Nature 560:253–257

Baell JB (2021) Personal accounts of Australian drug discovery at the public-private interface. Aust J Chem 74:16–27

Begley DW, Fox D, 3rd, Jenner D et al. (2014) A structural biology approach enables the development of antimicrobials targeting bacterial immunophilins. Antimicrob Agents Chemother 58 (3):1458–1467

Benkendorff K (2013) Natural product research in the Australian marine invertebrate *Dicathais orbita*. Mar Drugs 11:1370–1398

Benkendorff K, Bremner JB, Davis AR (2000) Tyrian purple precursors in the egg masses of the Australian muricid, *Dicathais orbita*: A possible defensive role. J Chem Ecol 26(4):1037–1050

Biswas NN, Iskander GM, Mielczarek M et al. (2018) Alkyne-substituted fimbrolide analogues as novel bacterial quorum-sensing inhibitors. Aust J Chem 71 (9):708–715

Blount KF, Breaker RR (2006) Riboswitches as antibacterial drug targets. Nat Biotechnol 24 (12):1558–1564

Boström J, Brown DG, Young RJ et al. (2018) Expanding the medicinal chemistry synthetic toolbox. Nat Rev Drug Discov 17:709–727

Boyaci H, Chen J, Lilic M et al. (2018) Fidaxomicin jams *Mycobacterium tuberculosis* RNA polymerase motions needed for initiation via RbpA contacts. Elife 7:e34823

Branson TR, Turnbull WB (2013) Bacterial toxin inhibitors based on multivalent scaffolds. Chem Soc Rev 42 (11):4613–4622

Brown DG, Boström J (2016) Analysis of past and present synthetic methodologies on medicinal chemistry: where have all the new reactions gone? J Med Chem 59 (10):4443–4458

Brüssow H (2017) Infection therapy: the problem of drug resistance - and possible solutions. Microb Biotechnol 10 (5):1041–1046

Buonanno F, Anesi A, Guella G et al. (2017) Blepharismins used for chemical defense in two ciliate species of the genus *Blepharisma*, *B. stoltei* and *B. undulans* (Ciliophora: Heterotrichida). Eur Zool J 84(1):402–409

Buroni S, Scoffone VC, Fumagalli M et al. (2018) Investigating the mechanism of action of diketopiperazines inhibitors of the *Burkholderia cenocepacia* quorum sensing synthase CepI: A site-directed mutagenesis study. Front Pharmacol 9: Article 836

Calvert MB, Jumde VR, Titz A (2018) Pathoblockers or antivirulence drugs as a new option for the treatment of bacterial infections. Beilstein J Org Chem 14:2607–2617

Carney DW, Schmitz KR, Truong JV et al. (2014) Restriction of the conformational dynamics of the cyclic acyldepsipeptide antibiotics improves their antibacterial activity. J Am Chem Soc 136:1922–1929
Chen H, Engkvist O (2019) Has drug design augmented by artificial intelligence become a reality? Trends Pharmacol Sci 40 (11):806–809
Chu J, Vila-Farres X, Inoyama D et al. (2016) Discovery of MRSA active antibiotics using primary sequence from the human microbiome. Nat Chem Biol 12 (12):1004–1006
Cockburn CL, Green RS, Damle SR et al. (2019) Functional inhibition of acid sphingomyelinase disrupts infection by intracellular bacterial pathogens. Life Sci Alliance 2 (2):e201800292
Cohen F, Aggen JB, Andrews LD et al. (2019) Optimization of LpxC inhibitors for antibacterial activity and cardiovascular safety. ChemMedChem 14:1560–1572
Cully M (2018) Roche taps potential antibiotics mine with Warp Drive Bio. Nat Rev Drug Discov 17:8
Culp EJ, Waglechner N, Wang W et al. (2020) Evolution-guided discovery of antibiotics that inhibit peptidoglycan remodelling. Nature 578 (7796):582–587
Duncan LF, Wang G, Ilyichova OV et al. (2019) The fragment-based development of a benzofuran hit as a new class of *Escherichia coli* DsbA inhibitors. Molecules 24 (20):3756
Elgart M, Soen Y (2018) Microbiome-germline interactions and their transgenerational implications. BioEssays 40 (4):1700018
Elnaas AR, Grice D, Han J et al. (2020) Discovery of a natural product that binds to the *Mycobacterium tuberculosis* protein Rv1466 using native mass spectrometry. Molecules 25 (10):2384
Feigman MS, Kim S, Pidgeon SE et al. (2018) Synthetic immunotherapeutics against Gram-negative pathogens. Cell Chem Biol 25 (10):1185–1194.e1185
Fer MJ, Corre LL, Pietrancosta N et al. (2018) Bacterial transferase MraY, a source of inspiration towards new antibiotics. Curr Med Chem 25 (42):6013–6029
Feringa BL (2017) The art of building small: From molecular switches to motors (Nobel Lecture). Angew Chem Int Ed 56 (37):11060–11078
Fischer N, Neumann P, Konevega AL et al. (2015) Structure of the *E. coli* ribosome–EF-Tu complex at <3 Å resolution by Cs-corrected cryo-EM. Nature 520:567–570
Frapwell CJ, Skipp PJ, Howlin RP et al. (2020) Antimicrobial activity of the quinoline derivative HT61 against *Staphylococcus aureus* biofilms. Antimicrob Agents Chemother 64 (5):e02073–02019
Geng J, Weishuo L, Zhang Y et al. (2019) Radical polymerization inside living cells. Nat Chem 11:578–586
Gerry CJ and Schreiber SL (2020) Recent achievements and current trajectories of diversity-oriented synthesis. Curr Opin Chem Biol 56: 1–9
Giddings L-A, Newman DJ (2015) Bioactive compounds from terrestrial extremophiles. Springer, Cham, Switzerland, pp 1–90
González A, Fillat MF, Lanas Á (2018) Transcriptional regulators: valuable targets for novel antibacterial strategies. Future Med Chem 10 (5):541–560
Gopal P, Dick T (2020) Targeted protein degration in antibacterial drug discovery? Prog Biophys Mol Biol 152: 10–14
Gu S, Cui D, Chen X et al. (2018) PROTACs: An emerging targeting technique for protein degradation in drug discovery. Bioessays 40 (4):e1700247
Gunderson CW, Segall AM (2006) DNA repair, a novel antibacterial target: Holliday junction-trapping peptides induce DNA damage and chromosome segregation defects. Mol Microbiol 59 (4):1129–1148
Haque S, Ahmad F, Dar SA et al. (2018) Developments in strategies for quorum sensing virulence factor inhibition to combat bacterial drug resistance. Microb Pathog 121:293–302
Hart EM, Mitchell AM, Konovalova A et al. (2019) A small-molecule inhibitor of BamA impervious to efflux and the outer membrane permeability barrier. Proc Natl Acad Sci 116 (43):21748

Heras B, Scanlon MJ, Martin JL (2015) Targeting virulence not viability in the search for future antibacterials. Br J Clin Pharmacol 79 (2):208–215

Herzik MA, Jr. (2020) Cryo-electron microscopy reaches atomic resolution. Nature 587 (7832):39–40

Ho H, Miu A, Alexander MK et al. (2018) Structural basis for dual-mode inhibition of the ABC transporter MsbA. Nature 557 (7704):196–201

Hu LI, Lima BP, Wolfe AJ (2010a) Bacterial protein acetylation: the dawning of a new age. Mol Microbiol 77 (1):15–21

Hu Y, Shamaei-Tousi A, Liu Y et al. (2010b) A new approach for the discovery of antibiotics by targeting non-multiplying bacteria: A novel topical antibiotic for staphylococcal infections. PLoS One 5 (7):e11818

Ikeya T, Hanashima T, Hosoya S et al. (2016) Improved in-cell structure determination of proteins at near-physiological concentration. Sci Rep 6 (1):38312

Imai Y, Meyer KJ, Iinishi A et al. (2019) A new antibiotic selectively kills Gram-negative pathogens. Nature 576 (7787):459–464

Jia X, Lynch A, Huang Y et al. (2019) Anthropogenic biases in chemical reaction data hinder exploratory inorganic synthesis. Nature 573 (7773):251–255

Jacobsson S, Mason C, Khan N et al. (2019) In vitro activity of the novel oral antimicrobial SMT-571, with a new mechanism of action, against MDR and XDR *Neisseria gonorrhoeae*: future treatment options for gonorrhoea? J Antimicrob Chemother 74: 1591–1594

Johnson EO, LaVerriere E, Office E et al. (2019) Large-scale chemical-genetics yields new *M. tuberculosis* inhibitor classes. Nature 571 (7763):72–78

Keller MD, Ching KL, Liang FX et al. (2020) Decoy exosomes provide protection against bacterial toxins. Nature 579 (7798):260–264

Koppermann S, Cui Z, Fischer PD et al. (2018) Insights into the target interaction of naturally occurring muraymycin nucleoside antibiotics. ChemMedChem 13 (8):779–784

Krajewski SS, Ignatov D, Johansson J (2017) Two are better than one: Dual targeting of riboswitches by metabolite analogs. Cell Chem Biol 24 (5):535–537

Kroeck KG, Sacco MD, Smith EW et al. (2019) Discovery of dual-activity small-molecule ligands of *Pseudomonas aeruginosa* LpxA and LpxD using SPR and X-ray crystallography. Sci Rep 9 (1):15450

Kumar V, Turnbull WB (2018) Carbohydrate inhibitors of cholera toxin. Beilstein J Org Chem 14:484–498

Leaver DJ, Dawson RM, White JM et al. (2011) Synthesis of 1,2,3-triazole linked galactopyranosides and evaluation of cholera toxin inhibition. Org Biomol Chem 9 (24):8465–8474

Lebraud H, Wright DJ, Johnson CN et al. (2016) Protein degradation by in-cell self-assembly of proteolysis targeting chimeras. ACS Cent Sci 2 (12):927–934

Lewis K (2013) Platforms for antibiotic discovery. Nat Rev Drug Discov 12 (5):371–387

Liu F, Ma S (2018) Recent process in the inhibitors of UDP-3-O-(R-3-hydroxyacyl)-N-acetylglucosamine deacetylase (LpxC) against Gram-negative bacteria. Mini Rev Med Chem 18 (4):310–323

Lobban CS, Hallam SJ, Mukherjee P et al. (2007) Photophysics and multifunctionality of hypericin-like pigments in heterotrich ciliates: A phylogenetic perspective. Photochem Photobiol 83:1074–1094

Lok C (2015) Mining the microbial dark matter. Nature 522 (7556):270–273

Luechtefeld T, Marsh D, Rowlands C et al. (2018) Machine learning of toxicological big data enables read-across structure activity relationships (RASAR) outperforming animal test reproducibility. Toxicol Sci 165 (1):198–212

Lupoli TJ, Vaubourgeix J, Burns-Huang K et al. (2018) Targeting the proteostasis network for mycobacterial drug discovery. ACS Infect Dis 4:478–498

Luther A, Urfer M, Zahn M et al. (2019) Chimeric peptidomimetic antibiotics against Gram-negative bacteria. Nature 576 (7787):452–458

Maeda T, García-Contreras R, Pu M et al. (2012) Quorum quenching quandary: resistance to antivirulence compounds. ISME J 6:493–501

Mashalidis EH, Kaeser B, Terasawa Y et al. (2019) Chemical logic of MraY inhibition by antibacterial nucleoside natural products. Nat Commun 10 (1):2917

Megget K (2018) Novartis exit from antibiotics a setback for race against resistance. Chem World 15(9):48–49

Monserrat-Martinez A, Gambin Y, Sierecki E (2019) Thinking outside the bug: Molecular targets and strategies to overcome antibiotic resistance. Int J Mol Sci 20 (6):1255

Mullard A (2017) The drug-maker's guide to the galaxy. Nature 549 (7673):445–447

Nickerson NN, Jao CC, Xu Y et al. (2018) A Novel inhibitor of the LolCDE ABC transporter essential for lipoprotein trafficking in Gram-negative bacteria. Antimicrob Agents Chemother 62 (4):e02151–02117

Nizalapur S, Kimyon Ö, Biswas NN et al. (2016) Design, synthesis and evaluation of *N*-aryl-glyoxamide derivatives as structurally novel bacterial quorum sensing inhibitors. Org Biomol Chem 14 (2):680–693

Nuti DE, Crump RB, Dwi Handayani F et al. (2011) Identification of circulating bacterial antigens by in vivo microbial antigen discovery. mBio 2 (4):e00136–11

Pant B, Kato Y, Kumagai T et al. (1997) Blepharismin produced by a protozoan *Blepharisma* functions as an antibiotic effective against methicillin-resistant *Staphylococcus aureus*. FEMS Microbiol Lett 155 (1):67–71

Peeters SH, de Jonge MI (2018) For the greater good: Programmed cell death in bacterial communities. Microbiol Res 207:161–169

Petrelli D, Buonanno F, Vitali LA et al. (2012). Antimicrobial activity of the protozoan toxin climacostol and its derivatives. Biologia 67:525–529

Psonis JJ, Thanassi DG (2019) Therapeutic approaches targeting the assembly and function of chaperone-usher pili. EcoSal Plus 8 (2). https://doi.org/10.1128/ecosalplus.ESP-0033-2018

Rader C (2015) How to minimalize antibodies. Nature 518 (7537):38–39

Rasko DA, Moreira CG, Li de R et al. (2008) Targeting QseC signaling and virulence for antibiotic development. Science 321 (5892):1078–1080

Renaud J-P, Chari A, Ciferri C et al. (2018) Cryo-EM in drug discovery: achievements, limitations and prospects. Nat Rev Drug Discov 17 (7):471–492

Ribet D, Cossart P (2018) Ubiquitin, SUMO, and NEDD8: Key targets of bacterial pathogens. Trends Cell Biol 28 (11):926–940

Rideout MC, Boldt JL, Vahi-Ferguson G et al. (2011) Potent antimicrobial small molecules screened as inhibitors of tyrosine recombinases and Holliday junction-resolving enzymes. Mol Divers 15 (4):989

Robinson A, van Oijen AM (2013) Bacterial replication, transcription and translation: mechanistic insights from single-molecule biochemical studies. Nat Rev Microbiol 11 (5):303–315

Robinson JA (2019) Folded synthetic peptides and other molecules targeting outer membrane protein complexes in Gram-negative bacteria. Front Chem 7: Article 45

Rodrigues T, Werner M, Roth J et al. (2018) Machine intelligence decrypts β-lapachone as an allosteric 5-lipoxygenase inhibitor. Chem Sci 9 (34):6899–6903

Sader HS, Dale GE, Rhomberg PR et al. (2018) Antimicrobial activity of murepavadin tested against clinical isolates of *Pseudomonas aeruginosa* from the United States, Europe, and China. Antimicrob Agents Chemother 62 (7):e00311–00318

Sakakibara D, Sasaki A, Ikeya T et al. (2009) Protein structure determination in living cells by in-cell NMR spectroscopy. Nature 458 (7234):102–105

Scapin G, Potter CS, Carragher B (2018) Cryo-EM for small molecules discovery, design, understanding, and application. Cell Chem Biol 25 (11):1318–1325

Schaenzer AJ, Wlodarchak N, Drewry DH et al. (2017) A screen for kinase inhibitors identifies antimicrobial imidazopyridine aminofurazans as specific inhibitors of the *Listeria monocytogenes* PASTA kinase PrkA. J Biol Chem 292 (41):17037–17045

Schilling NA, Berscheid A, Schumacher J et al. (2019) Synthetic lugdunin analogues reveal essential structural motifs for antimicrobial action and proton translocation capability. Angew Chem Int Ed 58 (27):9234–9238

Schütz C, Empting M (2018) Targeting the *Pseudomonas* quinolone signal quorum sensing system for the discovery of novel anti-infective pathoblockers. Beilstein J Org Chem 14:2627–2645

Segler MHS, Preuss M, Waller MP (2018) Planning chemical syntheses with deep neural networks and symbolic AI. Nature 555 (7698):604–610

Shang Z, Winter JM, Kauffman CA et al. (2019) Salinipeptins: Integrated genomic and chemical approaches reveal unusual D-amino acid-containing ribosomally synthesized and post-translationally modified peptides (RiPPs) from a Great Salt Lake *Streptomyces* sp. ACS Chem Biol 14:415–425

Shlaes D (2018) What's ahead for antibiotics in 2019? 26 Dec 2018, https://www.acsh.org/news/2018/12/26/whats-ahead-antibiotics-2019-13685, accessed 12 April, 2019

Silver LL (2014) Antibacterials for any target. Nat Biotech 32 (11):1102–1104

Sittampalam GS, Rudnicki DD, Tagle DA et al. (2019) Mapping biologically active chemical space to accelerate drug discovery. Nat Rev Drug Discov 18 (2):83–84

Smith PA, Koehler MFT, Girgis HS et al. (2018) Optimized arylomycins are a new class of Gram-negative antibiotics. Nature 561 (7722):189–194

Sousa MC (2019) New antibiotics target the outer membrane of bacteria. Nature 576 (7787):389–390

Srinivas N, Jetter P, Ueberbacher BJ et al. (2010) Peptidomimetic antibiotics target outer-membrane biogenesis in *Pseudomonas aeruginosa*. Science 327 (5968):1010–1013

Stevens M, Abdeen S, Salim N et al. (2019) HSP60/10 chaperonin systems are inhibited by a variety of approved drugs, natural products, and known bioactive molecules. Bioorg Med Chem Lett 29(9):1106–1112

Styles MJ, Blackwell HE (2018) Non-native autoinducer analogs capable of modulating the SdiA quorum sensing receptor in *Salmonella enterica* serovar Typhimurium. Beilstein J Org Chem 14:2651–2664

Su C-C, Morgan CE, Kambakam S et al. (2019) Cryo-electron microscopy structure of an *Acinetobacter baumannii* multidrug efflux pump. mBio 10 (4):e01295–01219

Tanaka T, Ikeya T, Kamoshida H et al. (2019) High-resolution protein 3D structure determination in living eukaryotic cells. Angew Chem Int Ed 58 (22):7284–7288

Tanouchi Y, Lee AJ, Meredith H et al. (2013) Programmed cell death in bacteria and implications for antibiotic therapy. Trends Microbiol 21 (6):265–270

Terra L, Dyson PJ, Hitchings MD et al. (2018) A novel alkaliphilic *Streptomyces* inhibits ESKAPE pathogens. Front Microbiol 9:2458

Thaker MN, Wang W, Spanogiannopoulos P et al. (2013) Identifying producers of antibacterial compounds by screening for antibiotic resistance. Nat Biotechnol 31 (10):922–927

Vetterli S, Zerbe K, Müller M et al. (2018) Thanatin targets the intermembrane protein complex required for lipopolysaccharide transport in *Escherichia coli*. Sci Adv 4:eaau2634

Vickers CF, Silva APG, Chakraborty A et al. (2018) Structure-based design of MptpB inhibitors that reduce multidrug-resistant *Mycobacterium tuberculosis* survival and infection burden in vivo. J Med Chem 61 (18):8337–8352

Vu H, Pedro L, Mak T et al. (2018) Fragment-based screening of a natural product library against 62 potential malaria drug targets employing native mass spectrometry. ACS Infect Dis 4 (4):431–444

Walsh CT, Wencewicz TA (2014) Prospects for new antibiotics: a molecule-centered perspective. J Antibiotic 67 (1):7–22

Wanted: a reward for antibiotic development (2018). Nat Biotechnol 36 (7):555

Wu G-Y, Shi X, Phan H et al. (2020) Efficient self-assembly of heterometallic triangular necklace with strong antibacterial activity. Nat Commun 11 (1):3178

Yin Z, Wang Y, Whittell LR et al. (2014) DNA replication is the target for the antibacterial effects of nonsteroidal anti-inflammatory drugs. Chem Biol 21 (4):481–487

York A (2020) New drugs for the antibacterial pipeline? Nat Rev Microbiol 18 (2):61–61

Zhan J, Jia H, Semchenko EA et al. (2019) Self-derived structure-disrupting peptides targeting methionine aminopeptidase in pathogenic bacteria: a new strategy to generate antimicrobial peptides. FASEB J 33 (2):2095–2104

Zhong W, Koay A, Ngo A et al. (2019) Targeting the bacterial epitranscriptome for antibiotic development: Discovery of novel tRNA-(N1G37)methyltransferase (TrmD) inhibitors. ACS Infect Dis 5 (3):326–335

Zipperer A, Konnerth MC, Laux C et al. (2016) Human commensals producing a novel antibiotic impair pathogen colonization. Nature 535:511–516

Index

J. Bremner, *Multiple Action-Based Design Approaches to Antibacterials*,
https://doi.org/10.1007/978-981-16-0999-2

The manufacturer's authorised representative in the EU is Springer Nature Customer Service Centre GmbH, Europaplatz 3, 69115 Heidelberg, Germany. If you have any concerns regarding our products, please contact ProductSafety@springernature.com

Printed and bound by CPI Group (UK) Ltd, Croydon, CR0 4YY
30/09/2026
02225330-0005